Sears, Roebuck Book *of* Barns:

A Reprint of the 1919 Catalog

PREFACE BY REBECCA HUNTER AND DALE WOLICKI

Published in the United States by Rebecca Hunter, Elgin IL.

Photography Credits: All photographs by Rebecca Hunter unless otherwise specified.

ISBN 978-9762096-14

Manufactured in the United States of America

First Edition December 2005

Table of Contents

Preface i

Summary of Contents of Sears Modern Home Catalogs ix

Known Sears Catalogs for Structures other than Modern Homes x

Barn Preservation Resources xi

Bibliography for Mail Order Homes and Barns xii

1919 Book of Barns 1

Acknowledgements

Without the expertise and assistance from the following, this book would not have come into being.

Mr. and Mrs. Marvin Dillon of Watseka, Illinois, for permission to photograph their barn.

The Galena Gazette, Galena, Illinois, for allowing reproduction of their photographs of the Sears "Cyclone" barn in Elizabeth, Illinois.

Linda Martinie Hoffmann, for proofreading the introduction.

Sara Hoffmann, for cheerfully accompanying me while I photograph barns.

Marilyn Markle, for her impeccable sense of design and infinite patience.

Arlene May, and the Sears Archives, for making historic materials available to the editors.

Ron Mensching, for making available photographs of his family's Sears Barn.

Marilyn Raschka, for her willingness to drop everything to go look at yet another barn.

Larry and Pat Stein of Sodus, Michigan, for permission to photograph their Sears barn.

The Honorable Helen Marie Taylor for permission to photograph her five barns in Orange, Virginia.

Mr. John Worthington, Michigan Barn Preservations Network, for providing updated information on barn preservation resources.

Preface

(top photo) This Sears barn, the "Country Gentleman," stands on the outskirts of Watseka, Illinois.

(bottom photo) The "Country Gentleman" as shown in the 1924 Sears Barns *Catalog.*

Our recent nostalgia for barns is reflected in paintings of agricultural buildings, picture calendars featuring barns, and books such as *The American Book of Barns*. This nostalgia takes us back to the years prior to 1920, when the largest segment of the United States population was rural. As urban areas expand, barns and other agricultural buildings are falling prey to developers. Most are demolished, but some are preserved by adaptive reuse as homes or commercial enterprises. Salvaged barn wood has become a valuable commodity for architects, interior designers and artists.

When Europeans first came to this country and began to spread westward, by necessity, every pioneer became a farmer. In order to live, most pioneers became subsistence farmers, growing crops and raising livestock for personal use. Most Americans whose great grandparents came to this country from across the ocean have forebears who farmed.

The largest building on the farm was the barn. It was used for the storage of farm products, for feed storage and to house animals. The English word "barn" originates from the Old French "bereaern," from "bere" (barley) and aern (place). The most common exterior finish for barns was horizontal or vertical wood siding.

Different types of barns evolved to support different functions. The earliest barns were built around a threshing floor, with mows on either side for feed storage.[1] In later barns, typically, housing for animals was located on the lower level, while feed was stored above in a haymow. Most barns had a wide entrance for farm equipment and large animals, and smaller doors for pedestrians. Windows provided light, and ventilators were located on the roof. Interior partitions for stalls, mangers and implements varied according to the type of stock and equipment to be housed.

Early barns were often simple gabled structures. Lean-to structures could easily be attached to provide additional indoor space. Gambrel roofed barns became the most common, although more costly to construct, because they afforded a much larger feed storage space under the roof.[2]

The move from subsistence farming to large scale commercial farming during the late nineteenth century also demanded greater barn capacity.[3] Round barns provided for efficient feeding of livestock in stalls arranged radially around a central silo. In New England, "continuous" barns, that is structures connected to one another, were commonly built to facilitate movement from one building to the next in inclement weather.[4]

Horse barns contained stalls, a tack room, granary, and a room for wagons and implements. The entrance was large enough to drive into with a horse team and wagon. Dairy barns typically had a concrete foundation raised four feet above grade level. Windows had ventilation shields to protect the stock from drafts. Space for milk storage, processing and cooling could be included in the barn, or put in a separate milk house. Stalls and milking stanchions were arranged according to the needs of the farmer.

Historical Background of Sears, Roebuck as a Supplier of Mail Order Goods

In 1861, only 14% of the United States population lived west of the Mississippi River; by 1890 this figure had increased to 27%.[5] Manufacturing facilities, however, remained concentrated in the eastern part of the country. It was at this time that the first mail order companies were founded, in order to provide manufactured goods to rural areas.

A key to the marketing of goods by mail was the expansion of transportation facilities and of rural postal delivery services, allowing for the possibility of moving manufactured goods from one location to another in a speedy and cost-effective manner.[6] Sears Roebuck and Company, founded in 1886 by 22-year-old Richard Warren Sears, was one of the earliest such companies, following the lead of Montgomery Ward, who founded his mail order catalog company in

---

[1] *An Age of Barns*, author's note.
[2] *A Barn Sampler*, p. 14
[3] *Ibid.*, p.11
[4] *The Barns of Lake County*, p. 53
[5] *Catalogs and Counters*, p. 9
[6] *Catalogs and Counters*, p. 17

(top photo) The Sears "Alhambra" model was the most popular of Sears' two Spanish revival designs. This example is located in Highland Park, Illinois.

(bottom photo) The "Josephine" is a two-bedroom bungalow. This example is one of three "Josephines" in Elgin, Illinois.

1872.[7] These two companies, as well as smaller competitors, used mail order catalogs to market necessities and luxuries to the rural market, where retail stores were limited.

Among items offered in early Sears catalogs were food staples, clothing, tools, jewelry, musical instruments and building supplies. Since 1986, Sears' history as a supplier of precut mail order homes has received much publicity.

General Information on Precut Homes

These well designed, practical, homes were made of top quality materials. Lumber and hardware were purchased in bulk, and the structural elements were cut to exact size at the mill and shipped to the customer by rail. Parts for a typical home filled two box-cars, and included approximately 10,000 numbered pieces of lumber, kegs of nails and other hardware, roofing tar and shingles, paint and varnish, and blueprints and instructions on how to assemble the house. Manufacturers claimed the pre-cut system would save the builder up to 30% compared to the cost of standard building methods.

In the days before home power tools, precut homes represented an enormous saving in labor and materials for the home builder. Catalog prices typically included only the building materials; cost of the finished house, including the lot, the foundation, and construction labor was usually about double the catalog price. To promote their homes, companies placed advertisements in national magazines and newspapers in major cities.

These houses were usually not distinctive architectural designs, but copies of the most popular styles of the day. House designs were standardized to reduce waste in materials, but customers were encouraged to personalize their order by moving windows or doors, adding porches, fireplaces, sunrooms, window boxes, trellises, or built in cabinetry, and by selecting exterior finish and colors. Heating, plumbing and electric systems were available at additional cost.

Pre-cut housing thrived until after World War II, when government regulations, tract housing construction methods and increased popularity of prefabricated and mobile housing meant that most pre-cut housing companies could no longer compete financially. However, Aladdin, Sterling and Lewis Companies, all of Bay City Michigan, remained in business until the 1970's and 1980's.

Sears as a Supplier of Mail Order Buildings

Beginning in 1908, Sears entered the housing market with a catalog of house plans, following the lead of architects such as George F. Barber, Max L. Keith, and William A. Radford, who sold residential building plans through catalogs. Between 1908 and 1954, Sears Roebuck and Company marketed over 450 different models of homes and cottages by mail order catalog.

From 1908 to 1914 Sears sold only plans, with enough material of standard sizes to construct the building. In 1915 Sears began marketing precut, or kit, buildings, at the same time offering mortgage financing in hopes of attracting customers who did not have the cash to purchase homes or farm buildings from competitors such as Aladdin or Gordon-Van Tine. However, due to financial losses during the Depression, Sears discontinued mortgage financing in 1933, and continued to market precut homes of simplified design and lesser quality until 1940.

Sears sold an estimated 60-70,000 homes from 1908 to 1940, with the maximum sales in 1929. No sales records from Sears are extant today; the above estimates were based on figures published in the catalogs tempered with annual sales figures from the Modern Homes Division.

After World War II Sears again began marketing homes, offering a limited number of partially prefabricated models under the brand name "Homart" (from HOMan and ARThington streets where the main Chicago office was located.) These did not seem to have popular appeal, and were discontinued around 1951. From about 1953 until 1954, Sears sold at least one house model of concrete block finished with stucco.

---

[7] *Catalogs and Counters,* p. 24

The Barns That Sears Built

As early as 1911, four barn designs were added to the Modern Homes Catalog. By 1915, corn cribs, hog houses, chicken houses, a tool house, and a milk house were added to the line of barns.

In 1918, Sears issued its first specialty Honor Bilt Modern Farm Buildings catalog, *The Book of Barns.* Specialty catalogs for barns and other farm buildings continued to be published until 1930, after which time Sears appears to have discontinued selling agricultural buildings. Estimates on the number of barns sold are not available.

In keeping with Sears' policy of copying popular home styles, the barns and other farm buildings also reflected popular styles of the day. Over the years, seven basic styles of barns were sold. The most popular were barns with gambrel roofs (19 different models), followed by gabled roofs (7 models), and Gothic rounded roofs (5 models). One clipped gable model was sold, from 1915-17 as plans and materials (model #17 or C17) and from 1918-29 as a precut (model #3009, the "Star"). One small cross gable model (#18) was sold from 1915-17. One octagonal model was offered from 1916-17 (#65) and as a precut in 1918 (#2085, the "Octagon"). A single round barn was offered only in the 1917 catalog (#2071).

The popularity of a given model can be judged by the total number of years it appeared in the catalogs, assuming that less commonly ordered models were soon discontinued. Therefore, in the field, we would expect to see a larger number of gambrel style Sears barns, and fewer of the gabled, Gothic, round and octagonal models. See the table on page iv for a listing of model numbers and/or names, and the years they were sold.

Barns offered from 1911 until 1917 were not precut; they were sold as plans and materials. Beginning with the 1918 specialty *Book of Barns,* farm buildings were offered as precuts. This parallels the construction methodology of Sears' residential buildings.

(top photo) Sears barn #2085, the "Octagon." once stood proudly in Roselle, Illinois on the Mensching farm. The land was sold to a developer and the barn was razed. (Photo courtesy of the Mensching family.)

(bottom left) The owner of the "Cyclone" barn outside Elizabeth, Illinois (see barn photo, right) and Galena historian Daryl Watson examine the original Sears blueprints for the barn. (Photo courtesy of The Galena Gazette, *Galena, Illinois.)*

(bottom right) This Elizabeth, Illinois barn, "The Cyclone," is one of the few Gothic-style Sears barns. (Photo courtesy of The Galena Gazette, *Galena, Illinois.)*

The interior vault of the Sears Gothic-style "Cyclone" barn in Elizabeth, Illinois (shown on page iii). The barn is included on page 9 of the 1919 Book of Barns. *(Photo courtesy of* The Galena Gazette, *Galena, Illinois.)*

The first precut homes were offered in 1915: five models out of the total of 109 models offered, increasing to 44 precut out of the total of 132 models in 1916, and 90 precuts in 1917.

Specialty barn and agricultural building catalogs were published by Sears from 1918 to 1930. During these years, such buildings were not included in Modern Homes catalogs. Many new barn models were introduced in 1918, along with the concept of precut barns. The 1919 *Book of Barns* reproduced in this book is nearly identical to the 1918 edition. The 1918 catalog has four additional pages of product advertising inserted near the end of the catalog; otherwise, the 1919 pages are exact copies of the 1918 pages.

The 1918 and 1919 catalogs contain the largest number of buildings offered in any Sears agricultural buildings catalog. These were the years before the population of the United States shifted from being primarily rural to primarily urban, and thus the years during which there was the greatest need for farm buildings.

In subsequent years, some older models were dropped, and a few additional models were offered, but the total number of designs continued to decrease as the country's need for agricultural buildings declined. The total number of barn models offered declined from a high of 26 in 1918 and 1919 to a total of only 13 in 1929.

The Great Depression took a heavy toll on the construction industry nationwide; housing starts were down 52% by 1932. The depression equally affected the construction of agricultural buildings. Perhaps in response to this, Sears apparently discontinued its agricultural buildings by 1931.

At right are data about barn models offered from 1911-1919, 1921, 1924, 1928 and 1929. As of publication of this reprint, other Sears barn catalogs were not available to the authors.

For a list of barns appearing in Sears Modern Homes Catalogs, see page ix.

NON-PRECUT MODELS

YEAR		1911	1912	1913	1914	1915	1916	1917
Model	Style							
11	gambrel							=C11
12	gambrel							=C12
14	gabled							
16	gambrel							=C16
17	clipped gable = 3009							=C17
18	cross gable							=C18
60	gambrel							
65	octagonal = 2085							=C65
66	gambrel							=C66

Marketing Sears Barns

Barns, as all other Sears products, were marketed primarily through mail order catalogs. Sears long relied upon publication of affidavits and testimonials from satisfied customers to encourage additional people to buy its products. This strategy was also used to promote barns. From 1914 to 1917, 27 testimonials from barn buyers were published in Sears Modern Homes catalogs. Testimonials came from Illinois, Indiana, Iowa, Michigan, Nebraska, Ohio, South Dakota and Wisconsin. For a complete listing of these testimonials, see Rebecca Hunter's book *Putting Sears Homes on the Map.*

In addition, Sears used cross marketing, advertising their residential buildings in the barn books (see page 52), and advertising their farm buildings in the Modern Homes catalogs.

PRECUT MODELS

YEAR Model	Style	Name	1918	1919	1921	1924	1928	1929
2054	gambrel	Pride of Farmstead	■	■	■			
2055	gambrel	Blue Ribbon	■	■	■	■	■	■
2056	gothic	Landmark			■		■	
2057	gambrel	Dairy Belle	■	■	■			
2058	gambrel	Dairy Queen	■	■	■			
2059	gambrel	Favorite	■	■	■	■	■	■
2060	gambrel	Ideal	■	■		■		
2061	gothic	Our Leader	■	■	■	■	■	■
2062	gambrel		■	■	■	■	■	■
2063	gambrel	Oneida	■	■	■	■	■	■
2064	gabled	Prairie Chief	■	■	■	■	■	■
2065	gabled	Gable Feeder	■	■	■			
2066	gambrel	Gambrel Feeder	■	■	■			
2067	gabled		■	■	■	■	■	■
2068	gambrel	Fairview	■	■	■	■	■	■
2071	round		■	■				
2085	octagonal	Octagon = 65	■	■				
3005	gabled	Pocahontas	■	■				
3006	gambrel	Colonial	■	■	■	■	■	■
3007	gambrel	Country Gentleman	■	■	■	■	■	■
3008	gabled	Pioneer	■	■	■	■		
3009	clipped	Star = 17	■	■	■	■	■	■
3010	gambrel	Keystone	■	■	■			
3011	gothic	Crescent	■	■	■	■	■	■
3022	gothic	Big Chief	■	■		■	■	
3023	gambrel	Hiawatha	■	■	■			
3023	gothic	Springfield						■
3024	gabled	Economical Feeder	■	■	■			
3024	gothic	Cyclone						■
3026	gothic	Cyclone	■	■	■	■	■	
3028	gabled	Greenfield					■	■
5296	gambrel					■		■
5297	gothic					■		
5298	gambrel	Economy				■	■	■
5299	gambrel	Stockdale					■	■

Construction of Sears Barns

Initially, barns were advertised as fixed dimensions, e.g. #60 was 62' by 32' in 1913-17. Barn materials were initially furnished as standard lengths of lumber, but in 1918, Sears offered its first precut barns, saving the buyer the money that would have been paid to a carpenter to measure and cut every board. In 1918, barn number 2068, the "Fairview" listed 48 options for size, ranging from 24' by 38' to 40' by 134'. Size commonly increased by four foot increments in width and six foot increments in length.

The foundations were not included in the materials package or in the precut kits. Farmers were advised to dig the foundation down to the permanent moisture line for their region. Farmers were also advised to check the state laws governing foundations.

Who Built Sears Barns

By 1918, with the advent of precut barns, specifications included pre-cut framing and doors ready to hang. The precut method "saves on expert labor," and Sears advertised that any handyman could erect the barns.

To erect a barn which was not precut, the farmer would most likely hire one or more carpenters. In order to erect a large barn in a reasonable amount of time, it is likely that a farmer would seek aid from neighboring farmers, as in the earlier barn-raising gatherings.

(top photo) This meticulously restored example of Modern Barn #3023, the "Springfield," is located on Mrs. Helen Marie Taylor's horse farm, Meadowfarm, in Orange, Virginia. Mrs. Taylor's restorations of her five Sears barns were honored by the National Trust for Historic Preservation. (Photo by Dale Patrick Wolicki, © 2005)

(bottom photo) The "Springfield" driveway barn as it appeared in the 1929 Sears specialty Barns catalog.

Gordon-VanTine Company's Gothic barn model # 201 is nearly identical to Sears barn "Our Leader" pictured on page 14 of the 1919 Sears Book of Barns. *Gordon-VanTine barns are often misidentified as Sears barns.*

Sears' Competitors in the Mail Order Agricultural Buildings Market

Sears was not the only mail order building supplier to offer farm buildings; of the nine known major suppliers of mail order homes, five are known to have offered agricultural buildings. Gordon-Van Tine and Montgomery Wards both preceeded Sears in the marketing of agricultural buildings. In 1909, Gordon-Van Tine offered six barns and six additional agricultural buildings. In 1910, Wards offered four barns, a poultry house and a city stable. Sears joined them in 1911, followed by Harris Brothers in 1912 and Aladdin by 1914.

Based on evidence from available catalogs, Harris Brothers and Aladdin abandoned the agricultural buildings market by 1919. Sears, Gordon-Van Tine and Wards went on to introduce precut barns in 1918, offering a variety of styles in varying lengths and widths. Wards discontinued their agricultural buildings by 1920, leaving Sears and Gordon-Van Tine to reign as sole suppliers of mail order barns until the early years of the Great Depression. The last known Gordon-Van Tine barns catalog is dated 1929, and the last known Sears version is from 1930.

Below are brief summaries of agricultural buildings offered by Sears' competitors. Dates listed after each company name are the years in the mail order housing business, not the years when agricultural buildngs were sold.

Aladdin Company, Bay City, Michigan 1906-1982

Aladdin Company appears to have originated the concept of the precut or "kit" home in 1906. Like their later competitors, they also produced agricultural structures. The 1914 Aladdin Houses catalog shows a gambrel barn and two gable design barns, and invites the reader to send for the special catalog on Farm Buildings. The 1917 catalog shows the same three Readi-Cut barns. The Special Barn is 16' by 24'; the other two are offered in 20 different combinations of length and width. Aladdin appears to have played a comparatively minor role in the production of farm buildings.

Gordon-Van Tine, Davenport Iowa 1908-46

As early as 1909, Gordon-Van Tine offered plans for agricultural buildings. Included in the 1909 catalog are two round barns, three gambrel barns, one cross gable barn, poultry houses, granary, ice house, corn crib, hog house and an implement shed. By 1912, agricultural buildings appeared in a separate catalog. Separate catalogs of agricultural buildings were published in 1918, 1923, 1926, 1929, and presumably also in the intervening years. The 1918 catalog is remarkably similar to the Sears 1919 edition in number and style of barn models, and in variety of other farm buildings. Testimonials with photographs reveal the locations of a number of Gordon-Van Tine barns.

Harris Brothers, Chicago, Illinois 1912-31

Beginning in 1912, Harris offered a variety of agricultural buildings in their house catalogs. The 1912 catalog includes one circle barn, eight gambrel barns, one clipped gable and one cross gable barn, and a corn crib. These, plus an additional cross gable barn are also included in the 1914 and 1915 catalogs. In 1916 and 1918, readers of the housing catalog desiring information on agricultural buildings are asked to send for the special Barns booklet. From 1920-28, no mention is made of agricultural buildings either in the house catalogs or in the building materials catalogs.

Montgomery Ward, Chicago, Illinois 1910-31 (Wardway Homes)

Wards, Sears biggest competitor in the mail order catalog market, was also a competitor in the market for agricultural buildings. Wards' 1910-1914 house catalogs offer four barns, a city stable for three horses, and a poultry house. From 1915-1916, four new barn models are offered. By 1918, barns proliferated in the Wards house catalogs. 16 variations of a stock barn appear, eight of these not pre-cut and eight "frame cut." 20 versions of a gable roof barn and 20 of a gambrel barn are included, each available either "frame cut" or not pre-cut. The different model numbers represent different barn lengths. The 1919 catalog offers the above, plus a grain crib and poultry and hog houses. From 1922-1931, there is no mention of agricultural buildings in the house catalogs, and to date, separate barn catalogs have not surfaced.

Locating Sears Barns Today

To discover and authenticate Sears barns today is one of the great challenges for architectural historians and researchers. The country's former rural areas are fast shrinking as population grows and as cities expand into areas which once were agricultural. This expansion usually entails the demolition of any farm buildings remaining on the land to be developed. Usually, these buildings have already fallen into disrepair. Because of this, the locations of most Sears barns are not known. Since Sears no longer has any sales records, there is no good resource for locating large numbers of Sears barns. Testimonials published in early Modern Homes catalogs list the locations of a small number of barns. Usually, the only information given is the city and state to which the buildings were shipped; the actual addresses have yet to be discovered.[8]

Data do not exist to allow us to determine what percentage of all barns built from 1909-1930 were mail order barns. If we may conclude that a mail order barn was about as common as a mail order home, we may extrapolate from available data on mail order homes.[9] Only an estimated 3% of housing starts in the 1920's are thought to be mail order homes; if the same is true for mail-order barns and farm outbuildings, they are indeed rare.

Authenticating Sears Barns

As is true of mail order homes, it is more difficult to authenticate mail order barns which were not precut. In these cases, we must rely on external appearance, interior and exterior measurements, surviving documents and oral history. Barns which were precut offer the possiblity of discovering part numbers on the basic structural boards. A "Cyclone" model on Mrs. Taylor's horse farm in Orange, VA and a "Cyclone" in Elizabeth, IL have the same part number format as Sears homes.

Below is the information currently known about part number location and format in homes from various mail order companies. It remians to be seen whether these data apply equally to agricultural buildings.

The wooden parts of a precut or "kit" building were numbered in order to facilitate construction. After the structure is built, it is usually possible to see some of these numbers. Presence of part numbers constitutes proof that the building is in fact a mail order kit. The style of the numbering may be a good indicator of which company built the house. Post-construction, the numbers are usually not visible on every board, so it may take a few minutes and a good flashlight to find one. Look on ceiling joists, attic rafters, stair risers and treads, and wall studs. In the case of barns, it is less likely that numbers will have been obliterated by plaster, wallboard or paint, so the chances of finding them should be increased.

1. SEARS part numbers are stamped on the wood in dark blue, black or gray ink. They are just over one inch high, and usually consist of a capital letter followed by one or more numerals, for example: A159, L23, C2. Rarely, the format is two numbers separated by a hyphen, for example 12-9. Very rarely, red ink was used. The numbers are usually found near the end of a board, on the wider surface. Homes built before 1915 and barns built before 1918 were not kits, and would therefore have no part numbers.

2 GORDON-VAN TINE/WARDS numbers are handwritten in grease pencil, and are usually found on the middle of a board. They consist of numerals, hyphenated in groups, e.g. 47-13-19, or four- or five-digit numbers. Part names are stenciled in capital letters (e.g. "ceiling joist" "top rail"). Delivery address may be stenciled in ink. Homes prior to 1916 were not precut.

3. ALADDIN Company numbers are handwritten in grease pencil, usually on the middle of a board. They consist of numerals, hyphenated in groups of two or three. Some of the terminal numbers are dimensional fractions, e.g. 42-18-11$\frac{3}{4}$. Part names may be stamped in ink. These companies produced precut homes from their inception, so older homes will also have numbers.

(top photo) The standard format for a Sears part number is a capital letter followed by one or more numerals, stamped near the end of a board.

(bottom photo) This part number in one of Mrs. Helen Marie Taylor's three "Cyclone" model barns in Orange, Virginia has the same format as most Sears precut house part numbers. (Photo by Dale Patrick Wolicki, © 2005)

---

[8] *Putting Sears Homes on the Map*, pp. 76-77

[9] *Elgin Illinois Sears House Research Project*, 1999

4. HARRIS Brothers numbers are stenciled in ink, often in the middle of a board, and may be numerals alone, or numerals and letters, or Roman numerals. E.g. 76, HR 50, RI 32 or AII. Model number and/or order number may be handwritten in grease pencil. Early homes and all barns were not precut.

For additional information about mail order buildings, refer to the bibliography on page xii.

Future Research

The authors hope that publication of this catalog reprint will reawaken interest in Sears barns, revive stories of their construction, and encourage owners of such building to carefully preserve them for the future.

Currently, efforts are underway in many states to compile listings of authenticated mail order homes from Sears and other companies. A similar effort would be valuable in order to locate as many barns as possible, before they fall prey to urban development, or are forgotten by relatives of the families who built them.

If you know of a barn or other agricultural building from Sears or any other mail order company, please notify Rebecca Hunter, at 903 Cedar Ave., Elgin IL 60120, or reach her by email at rebeccahunter1216@gmail.com. If possible, please provide pictures of all four sides of the building, exterior dimensions, and indicate whether part numbers were seen.

This 210' long barn on Mrs. Helen Marie Taylor's world renowned Egyptian Arabian horse farm in Orange, Virginia is not pictured in any Sears catalog available to the editors as of publication of this book. It may have been a custom model designed especially for Meadowfarm, since it was built during the years when Sears offered custom house designs. (Photo by Dale Patrick Wolicki, © 2005)

Summary of Contents of Sears Modern Homes Catalogs

The largest barn on Mrs. Taylor's horse farm, known as the "Sylvania" and pictured on the previous page, accomodates horse stalls and an indoor riding arena. (Photo by Dale Patrick Wolicki, © 2005)

1908 1st edition: 22 Modern Homes plans
1908 7th edition: 45 Modern Homes plans
1909: 73 Modern Homes plans
1910: (missing)
1911: 82 Modern Homes plans, 4 barn plans
1912: 78 Modern Homes plans, 4 barn plans
1913: 119 Modern Homes plans, Ready Made Houses and garages, 5 barn plans
1914: 84 Modern Homes plans, Ready Made Houses and garages, 6 barn plans, corn cribs, hog houses, chicken houses, silos
1915: 104 Modern Homes plans, 5 Cut and Fitted Modern Homes, 8 barn plans, corn cribs, hog houses, chicken houses, tool house, milk house, garages
1916: 88 Modern Homes plans, 44 Cut and Fitted Modern Homes, 7 barn plans, corn cribs, hog houses, chicken houses, tool house, milk house
1917 spring: 88 Modern Homes plans, 44 Cut and Fitted Modern Homes, 7 barn plans, corn cribs, hog houses, chicken houses, tool house, milk house
1917 fall: 90 Modern Homes, most cut and fitted; 9 Simplex sectional cottages, 4 Simplex garages, 12 barn plans, corn cribs, hog house, chicken house, tool house
1918: 90 Modern Homes, most cut and fitted; 9 Simplex sectional cottages, 4 Simplex garages
1919: 92 Modern Homes, most cut and fitted; 9 Simplex sectional cottages, 4 Simplex garages, 12 barns, corn cribs, hog house, chicken house, tool house
1920: 68 Modern Homes, most cut and fitted; Star garages, barns, Ready Made buildings
1921: 94 Modern Homes, most cut and fitted; portable buildings
1922: 86 Honor Bilt homes, most cut and fitted; 4 Econo Bilt homes, Simplex sectional portable buildings
1923: 79 Honor Bilt homes, Honor Bilt garages, 13 Simplex sectional summer cottages, Simplex garages, hunters' cabins, poultry houses
1924: 70 Honor Bilt homes, 8 Standard Bilt homes, Honor Bilt garages, 12 Simplex sectional summer cottages, Simplex garages, hunters' cabins, poultry houses, sun rooms
1925: 58 Honor Bilt homes, 8 Standard Bilt homes, sun rooms
1926: 73 Honor Bilt homes, 8 Standard Bilt homes, 8 Simplex sectional cottages, Simplex sectional garages
1927: 73 Honor Bilt homes, 8 Standard Bilt homes, 7 Ready Cut garages, 8 Simplex sectional cottages, Simplex garages
1928: 74 Honor Bilt homes, 8 Standard Bilt homes, 6 Ready Cut garages, 10 Ready Cut summer cottages, Simplex garages
1929: 76 Honor Bilt homes, already cut and fitted; 8 Standard Bilt homes, 6 Ready Cut garages, Ready Cut summer cottages, Simplex garages
1930: 88 Honor Bilt homes, already cut and fitted
1931: 68 Honor Bilt homes, already cut and fitted
1932: 68 Honor Bilt homes, already cut and fitted
1933 spring: 68 Honor Bilt homes, already cut and fitted
1933 fall: 84 Honor Bilt homes, already cut and fitted; garages, log cabins, summer cottages, building materials
1934: 84 Honor Bilt homes, already cut and fitted; garages, log cabins, summer cottages, building materials
1935: 48 Honor Bilt homes, already cut and fitted; garages
1936: 50 Honor Bilt homes, already cut and fitted; garages
1937: 50 Honor Bilt homes, already cut and fitted; garages
1938: 39 Honor Bilt homes, already cut and fitted; garages
1940: 39 Honor Bilt homes, already cut and fitted; garages
1941: Unbound, 14 Modern Homes building plans
1946: Homart Designs for Convenient Living; 25 home designs by Richard B Pollman. Ranch style homes, 900 - 1460 square feet. These homes were not designed or produced by Sears.
1949: Homart "Build Your Own Home at Low Cost"; 8 small sectional homes
1951: Homart "Here's the Sears Plan for you to Build Your Own Home at Low Cost"; 11 small sectional homes
1953-54: 1 small stucco covered concrete block home

Known Sears Catalogs for Structures other than Modern Homes

1916: Ready Made Buildings

1918: Honor Bilt Modern Farm Buildings

1918: Book of Barns, Honor Bilt - Already Cut

1919: Book of Barns, Honor Bilt - Already Cut

1921: Modern Farm Buildings Already Cut and Fitted

1924: Modern Farm Buildings Already Cut and Fitted, Barn equipment

1925: Simplex Sectional Garages and Summer Cottages

1927: Summer Cottages

1928: Modern Farm Buildings Already Cut and Fitted

1929: Modern Farm Buildings Already Cut and Fitted

1929: Summer Cottages

1930: Modern Farm Buildings Already Cut and Fitted

1930: Garage Catalog

1930: Summer Resort Cottages

1931: Summer Cottages

1934: Garages, Log Cabins, Plyhouses and Cottages

1935: Log Cabins and Cottages

1936: Log Cabins and Cottages

1937: Log Cabins and Cottages

1938: Log Cabins and Cottages

Barn Preservation Resources

(top photo) Is it a Sears barn, or not? This barn in Richmond, Illinois is very similar to Sears model #5298, but further investigation is needed before we can say it ***is*** *# 5298.*

(bottom photo) Sears Barn #5298 as pictured in the 1924 barns catalog.

BARN PRESERVATION SOCIETY

Dutch Barn Preservation Society, threeriversshms.com

BARN PRESERVATION PUBLICATIONS

Endersby, Elric; Larkin, David; Rocheleau, Paul; Greenwood, Alexander. *Barn: Preservation and Adaptation.* ISBN 0789307944

Endersby, Elric; Greenwood, Alexander; Larkin, David. *Barn: The Art of a Working Barn.* Boston: Houghton Mifflin Co., 1992

Giving Old Barns New Life Series; Volume 3: Bibliography on Barns and Barn Preservation. Wisconsin Historical Museum, Madison, WI 2005. 608 264-6555.

STATE BARN PRESERVATION ASSOCIATIONS

For states not listed below, contact the State Historic Preservation Office and/or the Cooperative Extension Office. Some counties may also have barn preservation organizations.

Colorado: Colorado Preservation Inc., 910 16th St, Suite 1100, Denver CO 80202

Illinois: Preservation Coordinator, Save Our Barns Committee, 328 S Douglas Ave, Springfield IL 62704

McHenry County Historical Barn Preservation Society, www.mchenrycountybarns.org

McHenry County Historical Society barn registration program, www.mchsonline.org

Indiana: Historic Landmarks Foundation of Indiana, 340 W Michigan, Indianapolis IN 46202

Iowa: Iowa Historic Preservation Alliance, PO Box 814, Mt Pleasant IA 52641

Maine: Historic Preservation Commission mainegov-images.informe.org

Michigan: Michigan Barn Preservation Network, PO Box 614, Mt Pleasant MI 48804-0614, www.mibarn.net

Minnesota: Preservation Outreach Coordinator, Minnesota Historical Society, 345 Kellogg Blvd, W St Paul MN 55102-1906

New York: Director of Technology and Grant Programs, League of New York State, 202 The Commons, Room 302, Ithaca NY 14850

New York State Barn Coalition www.barncoalition.com/barnpreservation.html

Ohio: Ohio Historic Preservation Office www.ohiohistory.org/resource/histpres/toolbox/barnagain.html

Friends of Ohio Barns, PO Box 203, Burbank OH 44214

Vermont: Division for Historic Preservation www.uvm.edu/~hag/barns/barncourse.html

Vermont Historic Preservation Program, University of Vermont, Burlington VT 05405

Wisconsin: Barn Preservation Program www.uwex.edu/lgc/barns.htm Community Planning and Design Specialist, Local Government Center, 610 Langdon St, Madison WI 53703

Barns NOW, 525 Vernon Ave, Madison WI 53703

Bibliography for Mail Order Homes and Barns

Sears Barn "The Hiawatha" in Sodus, MI, is one of the smaller barns offered by Sears.

Prepared by researcher Rebecca Hunter

BARNS

Arthur, Eric and Witney, Dudley. *The Barn: A Vanishing Landmark in North America.* Toronto: M F Feheley Arts Co., 1972

Korling, Diane *et al. A Barn Sampler: Rural Architecture in Kane County Illinois.* Unpublished ms., 1977

Leffingwell, Randy. *The American Barn.* St Paul MN: Motorbooks Classics, 2003. Prev. copyrights 1991 and 2002

Schumm-Burgess, Nancy. *The Barns of Lake County.* Virginia Beach VA: The Donning Co Publishers, 184 Business Park Dr, Suite 206, Virginia Beach VA 23462, 2004

Sloane, Eric. *An Age of Barns.* Warren CT: no date given

MAIL ORDER BUILDINGS

Books:

Emmet, Boris and Jeuck, John E. *Catalogs and Counters: A History of Sears, Roebuck and Company.* Chicago: University of Chicago Press, 1950

Fetters, Thomas. *The Lustron Home.* McFarland and Co, Box 611, Jefferson NC 28640, 2002

Hunter, Rebecca. *Beyond Sears: Elgin Illinois Mail Order Homes from Gordon-Van Tine, Aladdin, Lewis, Sterling, Montgomery Ward and Harris Brothers.* Available from the Gail Borden Public Library, Elgin IL; Rebecca Hunter, 2004

Hunter, Rebecca. *Elgin Sears House Research Project.* Available from the Gail Borden Public Library, Elgin IL; Rebecca Hunter,1999

Hunter, Rebecca. *Putting Sears Homes on the Map: A compilation of testimonials published in Sears Modern Homes Catalogs from 1908-1940.* Elgin IL; Rebecca Hunter, 2004

Reiff Daniel D. *Houses from Books.* University Park PA: Philadelphia State University Press, 2000

Schweitzer R. and Davis M.W. R. *America's Favorite Homes: Mail Order Catalogues as a Guide to Popular Early 20th Century Houses.* Detroit: Wayne State University Press, 1990

Stevenson K.C. and Jandl H.W. *Houses By Mail.* National Trust for Historic Preservation, 1986. Available from John Wiley & Sons Inc, 605 3rd Ave New York NY 10158-0012, 800 225-5945

Thornton, Rosemary. *The Houses that Sears Built.* Alton IL: Gentle Beam Publications, PO Box 1392 Alton IL 62002, 2002

Wolicki, Dale. *The Historic Architecture of Bay City, Michigan.* Bay City MI: The Bay County Historic Society, 1998

Catalog Reprints

Aladdin. *Aladdin "Built in a Day" House Catalog 1917.* Philadelphia and New York: Athenaeum and Dover Publications, 1995

Gordon-Van Tine. *117 House Designs of the Twenties.* Philadelphia and New York: Athenaeum and Dover Publications, 1992

Pacific Homes. *California's Kit Homes: A Reprint of the 1925 Pacific Ready-Cut Homes Catalog.* Alton IL: Gentle Beam Publications, PO Box 1392 Alton IL 62002, 2004.

Sears Roebuck. *Small Houses of the Twenties. The Sears Roebuck 1926 House Catalog.* Philadelphia and New York: Athenaeum and Dover Publications, 1991

Sears Roebuck. *Homes in a Box: Modern Homes from Sears, Roebuck and Co.* Schiffer Publishing, 1998. This is a reprint of the 1912 edition of the Sears Modern Homes catalog

Sears Roebuck. *Sears Roebuck and Co. Sears House Designs of the Thirties.* This is a reprint of the 1932 *Homes of Today* catalog. Philadelphia and New York: Athenaeum and Dover Publications, 2003

Sears Roebuck. *Sears, Roebuck Home Builder's Catalog 1910 Edition.* Philadelphia and New York: Athenaeum and Dover Publications, 1990

Montgomery Ward. *Wardway Homes, Bungalows and Cottages 1925.* Mineola New York: Dover Publications, 2004

ARTICLES

Billings, Mark. "In Praise of Kit Homes." *American Bungalow,* Spring 2003

Halpin, Kay. "Sears Roebuck's Best Kept Secret." *Historic Preservation,* Sept 1981

Hicks, L Wayne. "The House is in the Mail." *American History,* April 2000

Hunter, Rebecca. "Kit and Precut Homes: An American Architectural Phenomenon." *Preservation and Conservation Association Newsletter,* V. 23, No. 3-4, Champaign IL 2003

Maxwell, Shirley and Massey, James C. "Pre-Cut Houses." *Old House Journal,* Nov/Dec 1990

Murray, Alan. "Mail Order Homes Sears Sold in 1909-37 are Suddenly Chic." *The Wall Street Journal,* Feb 11, 1985

Maxwell, Shirley and Massey, James C. "The Story on Sears." *Old House Journal,* Aug 2002

Poore, Patricia. "Pattern Book Architecture: Is yours a mail order house?" *Old House Journal,* Dec 1980

Reiff, Daniel D. "Identifying Mail Order and Catalog Houses." *Old House Journal,* V. 5, 1995

Schwartz, David M. "When Home Sweet Home Was Just a Mailbox Away." *Smithsonian,* Nov 1985

Schweitzer, R. and Davis, M.W. R. "Aladdin's Magic catalog." *Michigan History*, Jan-Feb 1984

Snyder, Tim. "The Sears Pre-cut: A Mail Order House for Everyone." *Fine Homebuilding,* Aug-Sep 1985

Thornton, Rosemary. "The Sears Homes: Quality Homes Made to (Mail) Order." *Historic Illinois,* V. 25, No. 2. Illinois Historic Preservation Services, Aug 2002

Wolicki Dale. "Aladdin Homes: Comfortable, Convenient and Cozy." *American Bungalow,* Summer 2003

Wolicki, Dale. "Gordon-Van Tine Company." ickilow inc, 2002

ON THE WEB

www.kithouse.org: Services and publications by architectural historian and researcher Rebecca Hunter. Includes architectural salvage items.

www.searsarchives.com/homes: The website of Sears Roebuck Archives, including the national Sears Home Registry.

www.lib.cmich.edu/clarke/aladdin/catdir/htm: Clarke Library, University of Central Michigan, houses the original sales records of the Aladdin Company

group.yahoo.com/group/LustronHomes: information on Lustron homes; list of home locations.

ORIGINAL CATAOGS AND RECORDS

Aladdin, Lewis, Sterling Companies
Bay County Michigan Historical Society owns many of these catalogs. 321 Washington Av, Bay City MI 48708. 989 894-5733. Aladdin Company built examples of most of its homes in Bay City MI.

Clark Historical Library, Central Michigan University, Mount Pleasant MI, holds the Aladdin Company sales records. Contact Frank Boles, 989 774-3352

Gordon-Van Tine Company
Davenport IA Public Library has five catalogs published between 1926 and 1941, as well as three catalogs of building materials. 321 Main St, Davenport IA, 319 326-7832.

ORIGINAL CATAOGS AND RECORDS (CONT.)

Montgomery Ward
American Heritage Center, University of Wyoming, Laramie WY holds 14 home catalogs published between 1916 and 1931, and 17 building materials catalogs. 307 766-4114

Sears Roebuck
Sears Archives holds catalogs from 1908 to 1940, except 1910, 1911 and 1923. 847 286-9555, archive@sears.com, www.searsarchives.com

Multiple Companies
Gail Borden Library, Elgin IL 60120 has bound circulating photocopies of a number of catalogs from Sears, Gordon-Van Tine, Pacific Homes, and Harris Brothers. 847 742-2411

ORDER FORM

Putting Sears Homes on the Map: A Compilation of Testimonials Published in Sears Modern Homes Catalogs 1908-1940 is an important reference book for libraries, heritage commissions, historical societies and individuals. This study presents the only information available directly from Sears, Roebuck and Company about the locations of the homes they sold. Over 1300 entries tell us that Sears sold homes in at least 43 states.

Additional copies of *Sears Roebuck Book of Barns* and Hunter's *Putting Sears Homes On the Map* may be ordered using the form below.

Name: ____________________

Address: ____________________

City: ____________ State: ______ Zip code: ________

Sears Roebuck Book of Barns: # Copies: ______ x Price: $20.00 ea = ________

Putting Sears Homes On the Map: # Copies: ______ x Price: $20.00 ea = ________

Illinois residents add sales tax ________

Shipping $4.00 x # copies = ________

Total: ________

Make check or money order payable to Rebecca Hunter, and mail to:
REBECCA HUNTER
903 CEDAR AVE.
ELGIN IL 60120

or email: rebeccahunter1216@gmail.com

For further information, see www.kithouse.org

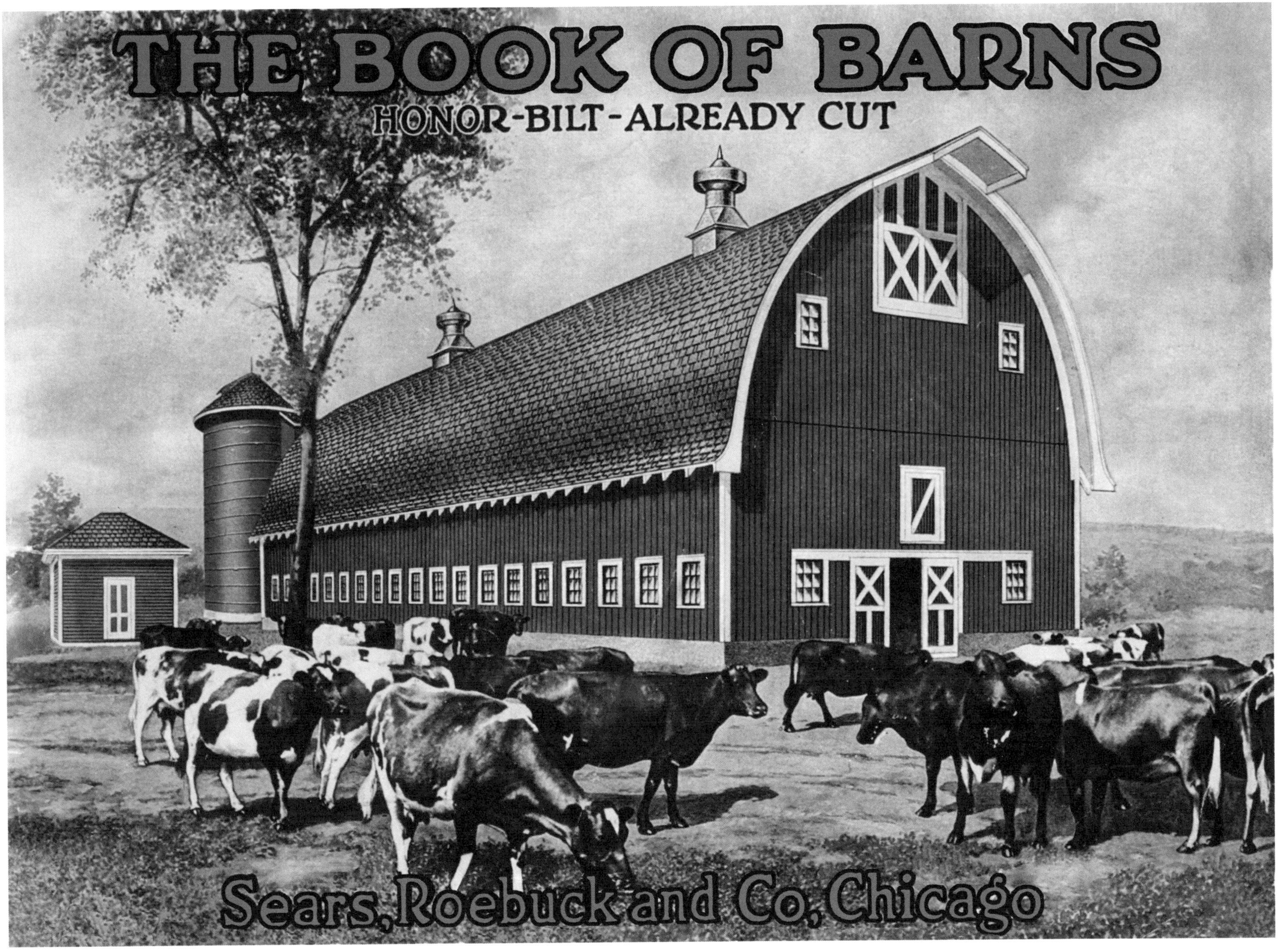
THE BOOK OF BARNS
HONOR-BILT-ALREADY CUT
Sears, Roebuck and Co, Chicago

GUARANTEE

We Guarantee that all of the material for our "Already Cut" Modern Farm Buildings will be cut and fitted, ready to nail in place.

We Guarantee the "Already Cut" Modern Farm Buildings described, illustrated, specified and priced in this catalog, as well as every other item listed and illustrated herein, to be exactly as represented.

We Guarantee that every Farm Building will arrive at destination in FIRST CLASS CONDITION.

We Guarantee to make PROMPT SHIPMENT of your order.

We Guarantee that after the shipment has been examined at the station, if the Farm Building does not come up to the purchaser's expectations in every respect, by notifying us in writing it can be shipped back AT OUR EXPENSE and we will promptly RETURN THE MONEY PAID and any freight charges that may have been paid by the purchaser.

We Guarantee to replace any piece or part that may be damaged in transit or otherwise considered unsatisfactory, or to return the price of such item, as the purchaser may direct, upon receipt of the paid freight expense bill with a notation by the freight agent stating the damage incurred.

If you comply with our simple requirements, we will send you an "Already Cut" Modern Farm Building costing $500.00 or upwards on EASY PAYMENTS. Orders for lower priced buildings must be accompanied with cash in full. Full particulars may be had on application.

Remember, a guarantee is worth only what is behind it. Sears, Roebuck and Co. stand back of this guarantee.

SEARS, ROEBUCK AND CO. **CHICAGO, ILLINOIS**

573R

MADE ENTIRELY FROM SELECTED LUMBER

BETTER FARM BUILDINGS

SHELTERED FROM RAIN, SUN, SOOT AND WIND

The Care We Take

In protecting our lumber enables us to guarantee the quality and insure complete satisfaction to our customers.

From the time the lumber leaves the sawmill until it is required in our factory it is thoroughly protected from the weather. Notice the 800-foot loading platform—one of ten—in the small picture and the large, well built rainproof warehouses below. Is it not reasonable that the doors, windows, frames, etc., manufactured from high grade lumber so protected should be more satisfactory in every way than the same articles made from lumber exposed to rain, snow, hail, soot and dust?

We load our lumber direct from our covered platforms into box cars for QUICK DELIVERY TO OUR CUSTOMERS. "Better lumber" means BETTER FARM BUILDINGS.

Methods That Make Better Material

If we could show you through our modern factory and let you see for yourself the many money saving devices we have installed, you would understand why we offer such excellent quality at the low prices printed in this book.

The high grade lumber which you see piled in the big warehouses below is always clean and dry. Each pile of lumber, containing nearly a carload, rests upon a movable truck which, when the lumber is needed, can be quickly transferred to the tracks in the foreground and pushed either into the dry kilns or wherever required. All lumber handled at our factory is unloaded from the cars on to these trucks and kept ready for use at a moment's notice.

You can see how different this process is from the old custom where selections were made from one pile and another and carried by wagon in small quantities (requiring much time and labor) to the different places around the factory where the workmen were waiting for it. All the money saved by our quick, modern system of handling lumber GOES DIRECT TO YOU in the way of better material for YOUR MODERN FARM BUILDING.

THESE SOLIDLY CONSTRUCTED RAINPROOF WAREHOUSES PROTECT OUR LUMBER UNTIL IT IS REQUIRED IN THIS FACTORY.

THIS ROOFING FOR YOUR MODERN HOME OR BARN.

APPROVED BY THE UNDERWRITERS' LABORATORIES INC. UNDER

LOOKS
COSTS
AND
OUT
WO
SHIN

ORIENTAL FIRE CHIEF SHINGLE ROLL

Our Own Trade Mark, Reg. in U. S. Pat. Off.

LOOKS LIKE PAINTED RED OR GREEN WOOD SHINGLES, but costs about half as much. Wears practically twice as long, and does not catch fire from sparks. Fire-Chief Shingle Roll Roofing does not have the defects, such as curled or warped shingles, so often found in wood shingle roofs.

THE PATTERN is made of red or sea green crushed slate (your choice of color), with black shaded lines inlaid between each shingle, a treatment that is permanent and beautiful. Needs no painting.

EITHER COLOR will make a nice roof, but we recommend the green because the shingle pattern is more pronounced.

EXCLUDES SNOW AND WATER BETTER THAN WOOD SHINGLES. Fire-Chief Shingle Roll Roofing is airtight and watertight.

WARMER THAN WOOD SHINGLES and helps to make a house cooler in Summer and warmer in Winter. Can be used over old shingles, giving double roof protection.

GUARANTEED FOR FIFTEEN YEARS. A written guarantee, backed by the entire resources of Sears, Roebuck and Co., protects you for fifteen years. If building is kept in good condition this roofing will wear many years longer.

48A3320—Red Fire-Chief Shingle Roll Roofing.
Price, per roll of 108 square feet.......................... **$3.75**

Shipped from CHICAGO, ILL., or SOUTHERN OHIO.

Shipped from either of the following warehouses: ALBANY, N. Y., ST. PAUL, MINN., KANSAS CITY, MO., or ATLANTA, GA., at 30 cents per roll extra.

For 1¾-in. nails to lay this roofing over old shingles, add 7c per roll extra.

386R—Barn Catalog.

SOLD ONLY BY SEARS, ROEBUCK AND CO., CHICAGO. FOR ORIENTAL RIDGE

SAFE, DURABLE, BEAUTIFUL AND ECONOMICAL

BETTER
LESS
WILL
WEAR
OD
GLES

DIRECTION OF THE NATIONAL BOARD OF FIRE UNDERWRITERS

SLATE SURFACED ROOFING

IT PAYS TO BUY GOOD ROOFING because the roof of a building has to stand the hardest weather conditions.

WEARING QUALITY FULLY TESTED. The formula and process for making Fire-Chief Shingle Roll Roofing have been in use for years in our Oriental Slate Surfaced Roofing. It has been put to the most severe tests. Roofing laid ten years ago is practically as good as the day it was laid. It is an asphalt roofing that defies heat, cold, rain and snow.

WE HAVE MADE THIS THE HEAVIEST, THICKEST, STRONGEST AND MOST BEAUTIFUL of all our roofings. It is the only natural effect shingle roll roofing in the market that we know of. Patented and sold only by us. Each roll weighs 90 pounds. **IT IS** the **HEAVIEST ROOFING WE HANDLE.**

THIS SHINGLE ROOFING IS FURNISHED IN ROLLS containing 108 square feet, 36 inches wide, with 8 square feet extra allowed for 2-inch laps, and will cover 100 square feet. All in one piece.

CEMENT AND GALVANIZED RUST RESISTING 1-INCH NAILS INCLUDED.

48A3321—Sea Green Fire-Chief Shingle Roll Roofing.
Price, per roll of 108 square feet.......... $3.75

Shipped from CHICAGO, ILL., or SOUTHERN OHIO.

Shipped from either of the following warehouses: ALBANY, N. Y., ST. PAUL, MINN., KANSAS CITY, MO., or ATLANTA, GA., at 30 cents per roll extra.

For 1¾-in. nails to lay this roofing over old shingles, add 7c per roll extra.

FIRE RESISTING

90 POUNDS

No 3321
GREEN

ORIENTAL
FIRE CHIEF
SHINGLE ROLL ROOFING

Guaranteed
15
YEARS
FOR WEAR AND COLOR

If the building on which it is laid remains in good condition, this roofing will wear many years longer.

UNDERWRITERS' LABORATORIES, INC.
INSPECTED
ROOFING MATERIAL
THIS LABEL MEANS
FIRE-RESISTING ROOFING

ROLL, FLASHING AND VALLEY TO MATCH THIS ROOFING. WRITE FOR PRICES.

LOOK AT THIS CLOSELY. EVERYTHING FOR YOUR

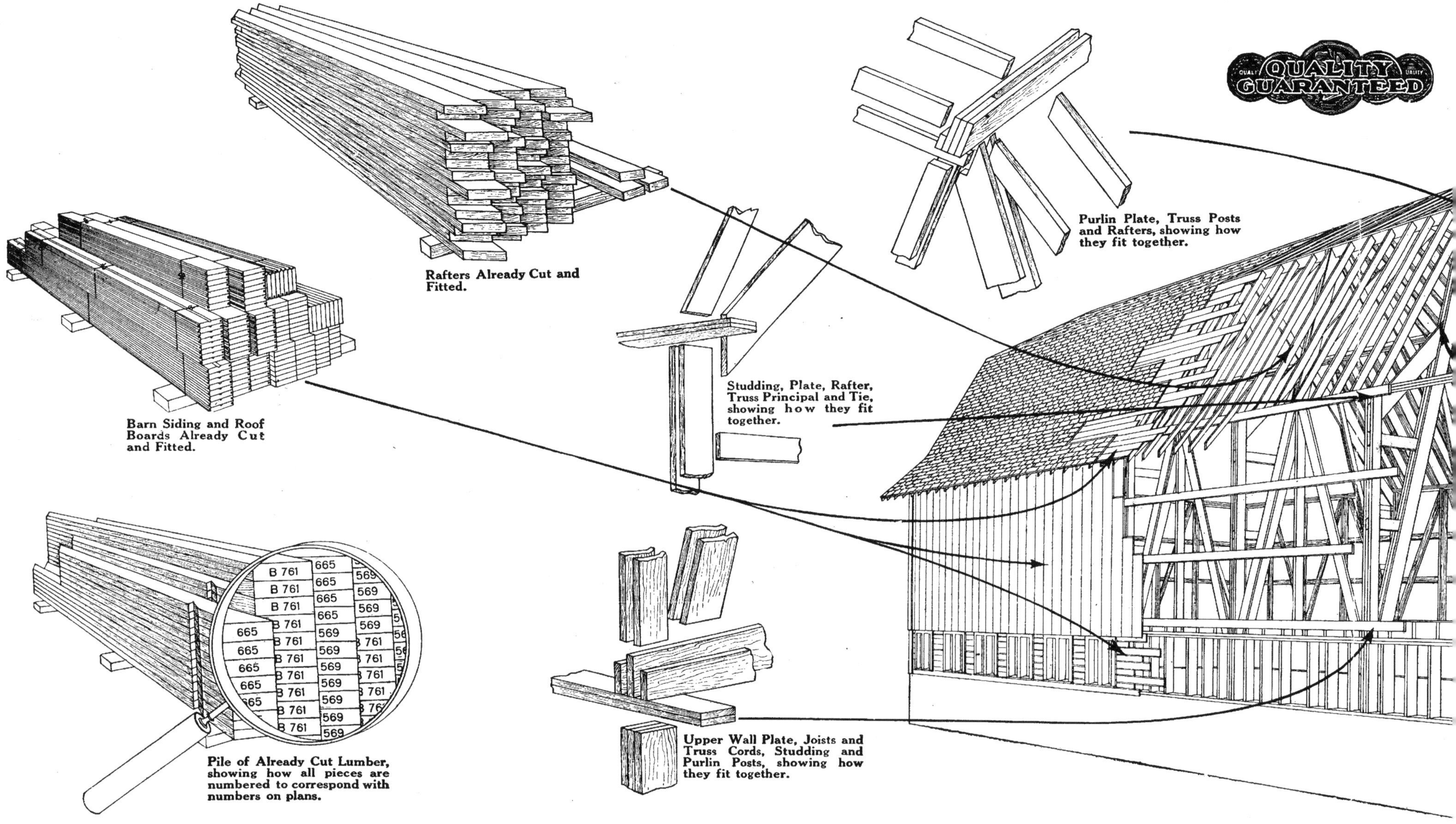

Rafters Already Cut and Fitted.

Barn Siding and Roof Boards Already Cut and Fitted.

Studding, Plate, Rafter, Truss Principal and Tie, showing how they fit together.

Purlin Plate, Truss Posts and Rafters, showing how they fit together.

Pile of Already Cut Lumber, showing how all pieces are numbered to correspond with numbers on plans.

Upper Wall Plate, Joists and Truss Cords, Studding and Purlin Posts, showing how they fit together.

SEARS, ROEBUCK AND CO. - -

386R

MODERN FARM BUILDING CUT AND FITTED

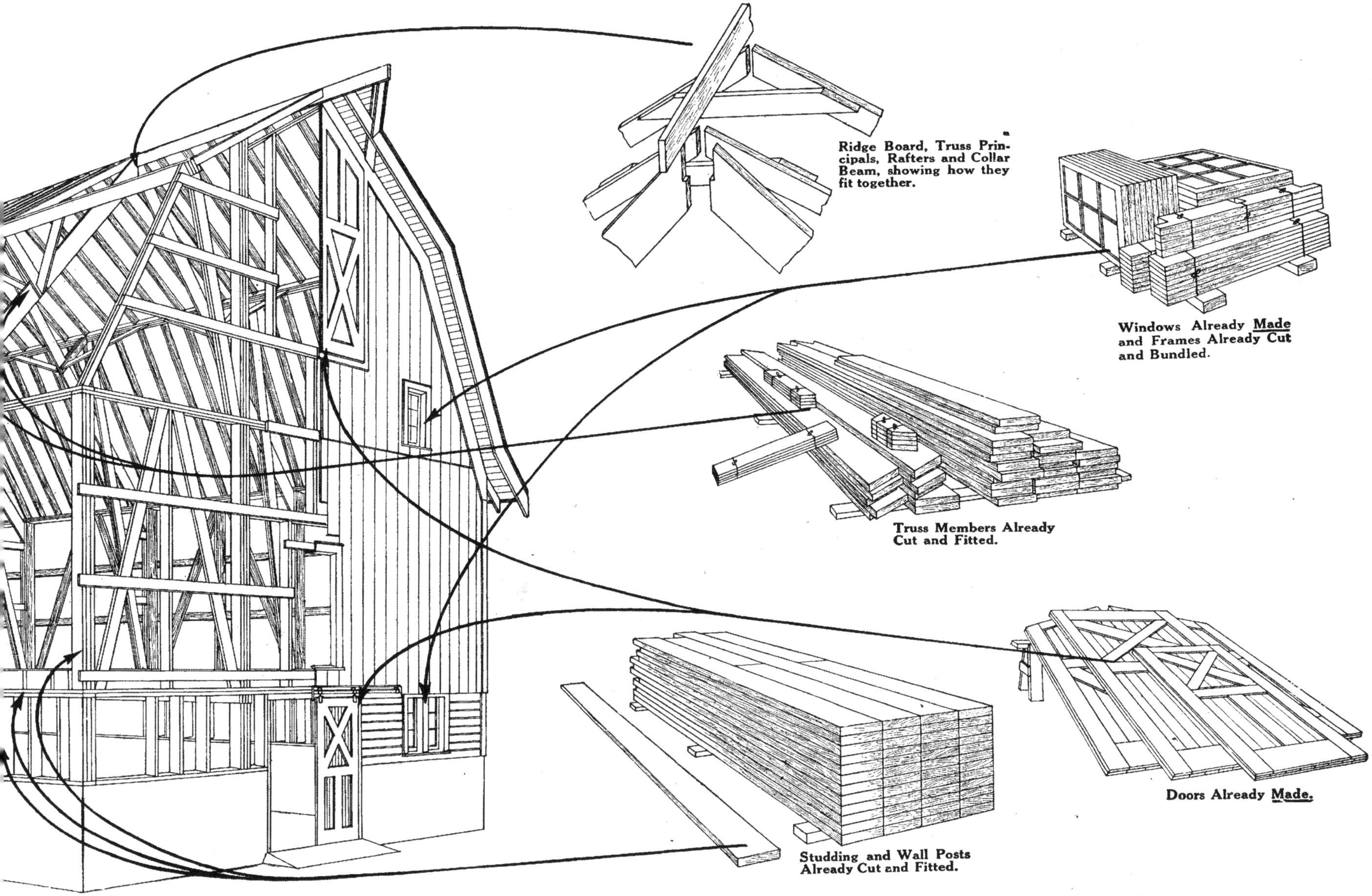

■ ■ ■ ■ CHICAGO

Eight Good Reasons Why You Should Buy Sears, Roebuck and Co.'s "Already Cut" Modern Farm Buildings

1. Our farm buildings are constructed scientifically. They combine the good features of the best farm buildings now in use.

2. Our farm buildings are designed according to the latest requirements of sanitation and convenience.

3. For every farm building the framing material is already cut and fitted. This prevents waste of material and reduces the labor about one-half. In addition, all doors are "ready made," ready to hang in place, making doors superior to those produced by hand carpentry.

4. Any handy man can erect our "already cut" modern farm buildings according to our plans. This saves a considerable amount on expert labor.

5. You have our guarantee that at the prices shown in this catalog we furnish enough material to complete the farm building. You will not be required to buy extra material.

6. We guarantee the material to be equal to or better than the grades specified in our general specifications.

7. When you deal with us you have the assurance that you are not paying for an experiment. We were the first to develop the mail order system of selling complete buildings. We have had many years of experience in this line. Our buildings are not experimental, but, having been built many times, have proved correct and practical in every detail.

8. You can buy any building in this book costing $500.00 or more on easy payments. For particulars see the Order Blank.

Notes About Foundation Walls of Barns

As a rule you may accept the state laws governing the depth of foundations for other buildings as a pretty sure indication of what is actually required. Some states say 24 inches; others, 30 inches; according to the severity of the Winter. But the frost line alone is not the controlling factor; other considerations such as loose or refilled ground merit consideration and should not be overlooked. It is always practical to go down to the permanent moisture line.

The height of the foundation wall above the established grade line merits your special attention because **your barn** and every other building **will last much longer if you prevent moisture from coming into contact with the sill and the framework of the building.**

In most of our barns we specify the height of the foundation above the grade line to be at least 24 inches. This has proved sufficient to meet ordinary requirements.

The illustrations on pages 4 and 5 convey a fairly good idea of the solid foundations and cross sections. They can be modified to meet particular requirements.

Cement or other foundation material is not included in the prices quoted on any of the buildings shown in this catalog.

The following proportions of cement, sand and crushed stone for concrete are recommended: One part of cement, two and one-half parts of sand, five parts of crushed stone **for foundation walls.** One part of cement, three parts of sand, six parts of crushed stone **for concrete floors in stables, mangers, etc.** One part of cement, one part of sand **for finishing floors, etc.**

As packed by the manufacturer, 1 barrel cement equals 4 sacks, each sack weighs 95 pounds and is equal to about 1 cubic foot.

For a cubic yard of solid concrete the following quantities of cement, sand and gravel or broken stone are required, proportioned as shown above.

	Cement, Cubic Feet	Sand, Cubic Feet	Gravel or Stone, Cubic Feet
Foundations	5	12½	25
Floors	4¼	12¾	25½
Finish	1	1	None

Variation in the fineness of sand and the solid tamping of the concrete may affect the quantities a little in either direction. 1 cubic yard of concrete will yield about 6 lineal feet of foundation wall as shown on page 4, which is 30 inches below and 24 inches above the grade line or about 5 lineal feet of foundation wall as shown on page 5, which is 30 inches below and 4 feet above the grade line, at a cost of approximately 35 and 40 cents per lineal foot for cement, figured on a price of $1.60 per barrel of cement.

All prices quoted in this book are for material loaded on the cars at our factory or yards.

Lumber and Mill Work for farm buildings are quoted loaded on the cars at our yards in SOUTHERN ILLINOIS.

Hardware, Paint and other Accessories are quoted loaded on the cars at CHICAGO, ILL.

Index to Barn Catalog

Page
Barns 6-30
Annex Barns 23, 26
Bank Barns 11, 16-17-18
Driveway Barns 19-20-21
Feeder Barns..........22, 24-25
Gothic Roof Barns.9, 14, 16-17, 27
L Shaped Barns............30
Round Barns, Etc.........28-29
Sheep Barns 26
Small Barns 27
Box Stalls 45
Corn Cribs...............35, 39
Doors 43-44
Eight Good Reasons..........2
Farm Buildings "Already Cut".3
Feed Racks 45
Floor Plans 40-41-42
Foundations, Etc.............2
General Specifications 4-5
Guarantee..Inside Front Cover
Hardware 50
Hog Houses.........31-32-33-34
The "Colony" Hog House...34
The "Community" Hog House 32
The "Monitor" Hog House..33
The "Sunlight" Hog House..31
The "Wigwam" Hog House.34
Ice Houses 38
Implement Sheds 39
Milk Houses 37
Modern Homes...............52
Paints 47
Partitions, Etc..............46
Paving Blocks...............51
Poultry Houses 36
Sheds 23
Silos 49
Ventilators 48
Windows 44

"ALREADY CUT" FARM BUILDINGS. REQUIRE NO SAWING. REDUCE LABOR ONE-HALF

No Waste—No Mistakes—No Big Labor Bills

READ THIS ABOUT OUR MACHINE MADE, "ALREADY CUT" MODERN FARM BUILDINGS.

This is the way your material comes when you buy one of our "Already Cut" Modern Farm Buildings:

All Wall Plates
All Girders
All Joists
All Studding
All Ridge Boards
All Plates
All Rafters
All Trusses

} ***"Already Cut" and Fitted, Ready for Nailing***

All Girts and Ribbons
All Braces
All Posts
All Door Frames
All Sash and Window Frames

} ***"Already Cut" and Fitted, Ready for Nailing***

All Barn Doors, ***ready made.***
All Windows and Sash, ***ready made.***

Each and every piece of the above material is **measured, fitted and cut to exact size for you at our factory.** The only work necessary on the job is to nail the pieces in place, following the instructions and plans furnished free with the material. Each piece is plainly marked to show just where it goes in the building. You can readily understand that when all this work is done at our factory, building is a small job. In addition, our price for a complete barn includes Window Shields with all six-light and nine-light windows for ground floor ventilation; fifteen-year guaranteed Fire-Chief Shingle Roll Roofing for all barns (except round barn), complete Hardware, Latches, Hinges, Staples, Nails, etc., and Paint, for two coats, your selection of colors.

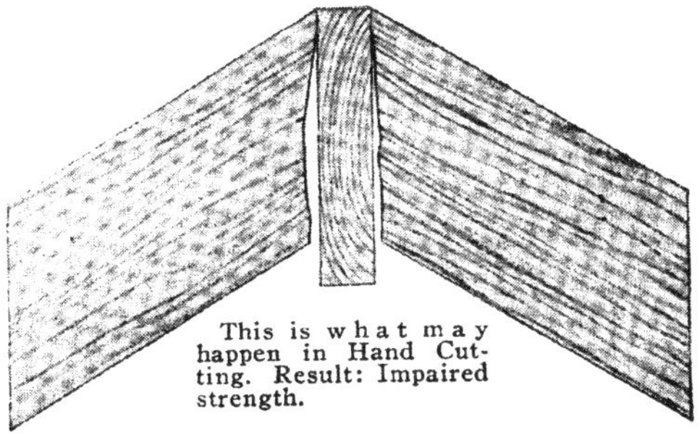

This is what may happen in Hand Cutting. Result: Impaired strength.

We furnish you all the Roof Boards, Barn Boards, Drop Siding, Mow Floor, Sub-Floors, as specified, in lengths that will work out to the best advantage on the job. When you purchase from us you know exactly what you get and how we cut and fit the material.

Our "Already Cut" buildings combine high quality of material, careful design and solid construction.

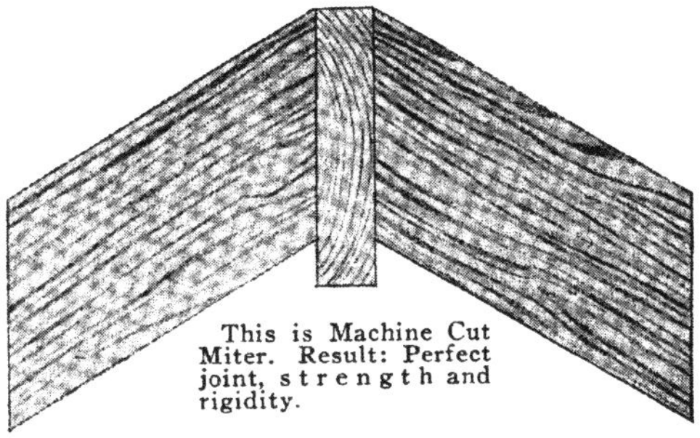

This is Machine Cut Miter. Result: Perfect joint, strength and rigidity.

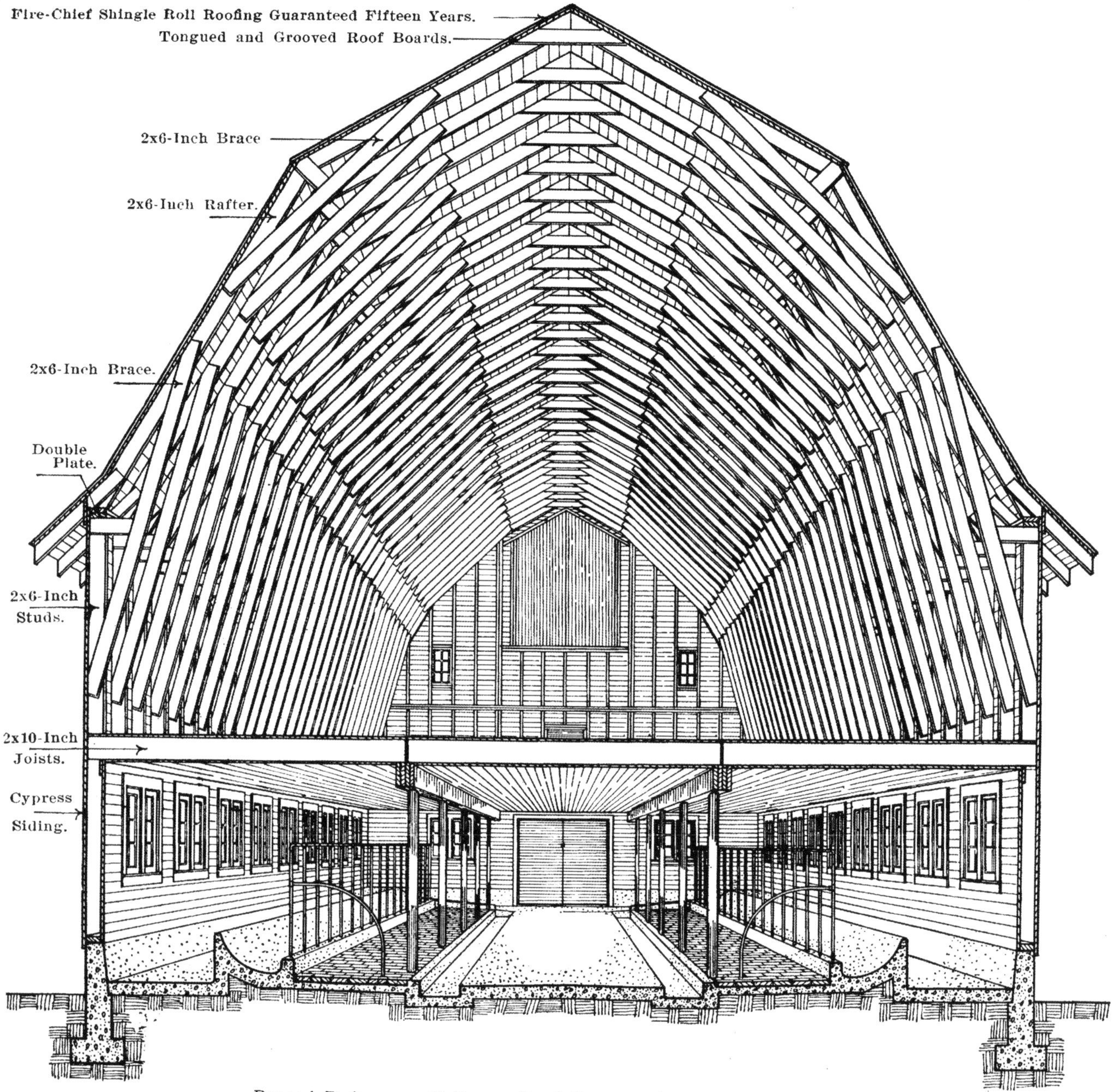

Braced Rafter—or Balloon—Roof Construction, for Barn 36 Feet Wide.
Regarding equipment, inside lining, ceiling, etc., see "Optional Items," page 5.

General Specifications

The illustrations on this and the following page are reproductions of the construction of most of our barns. No words are needed to describe the strength and rigidity of the construction. The illustrations speak for themselves. Equipment, inside lining and ceiling are optional; see page 5.

During our study of barn building we have carefully analyzed the methods of the builders and the requirements of the owners. We learned that three different types of construction are very popular—the **BRACED RAFTER CONSTRUCTION** or, as it is sometimes called, the Balloon Construction; the **TRUSSED ROOF CONSTRUCTION** and the **GOTHIC ROOF CONSTRUCTION.** The popularity of these types of construction is so great that we did not feel warranted in omitting any of them. On this and the following page we illustrate the Braced Rafter Construction and the Trussed Roof Construction. A good impression of the simplicity of the Gothic Roof Construction, its beauty and strength may be had by referring to the front and back covers of this book.

The same care used in selecting the most suitable types of construction was exercised when we decided on the kind of material to be specified. Prominent barn men agree that **THE FRAME OF THE BARN** should be constructed of the strongest lumber obtainable. Actual experience and statistics of the U. S. Department of Agriculture as well as many other sources prove that **YELLOW PINE IS THE MOST SUITABLE FOR THIS PURPOSE.** Therefore, we have decided to use only No. 1 Yellow Pine for the framework of all our barns in the belief that by this selection we meet actual requirements to the greatest advantage.

The quality of the **LUMBER TO BE USED FOR THE OUTSIDE WALLS** of a barn needs no less attention than the quality of the material used for the frame of the barn. But, while strength of material is the principal feature that governs the selection of the framing lumber, weather resisting qualities are the chief requirements in the selection of the lumber for the outside. We know of no higher quality of weather resisting lumber than that of **CYPRESS, "THE WOOD ETERNAL."** The superiority of cypress over other woods as a weather resisting material is well known. Bulletin No. 272 of the U. S. Department of Agriculture states that Cypress Farm Buildings in excellent preservation after 100 to 200 years and others of a like character are frequently reported.

Drop Siding.

GENERAL SPECIFICATIONS—Continued

We furnish two grades of Cypress Siding: **SELECT GRADE and No. 1 COMMON GRADE.**

Select Cypress Siding is our highest quality of Barn Siding and we recommend it highly. No. 1 Common Cypress Siding contains sound knots and other defects. It is favored in many localities for the reason that when it is painted it has the same appearance from the outside as the higher grade siding. Such defects in Cypress in no way impair the long life of the siding. In fact this grade will give double the length of service of even the highest grade of white pine.

A glance at the illustrations of "Drop Siding" and the "Double V Siding" will show you the designs we furnish for the outside walls of our barns. Double V Siding is mostly used on Truss Construction and Drop Siding on Braced Rafter Construction. Complete specifications will be found in the description of each barn. **FOR MOW FLOORING** we furnish **TONGUED AND GROOVED No. 1 YELLOW PINE.** This is superior to plain boards as it prevents dust from dropping on the stock below. A mow floor not airtight does not permit of good ventilation of the ground floor where the stock is kept. This is of greatest importance. In driveways through the mow floor of bank barns we lay **2-INCH PLANKS ON A TONGUED AND GROOVED YELLOW PINE SUB-FLOORING.** For the roof we furnish tongued and grooved, tight fitting roof boards of Yellow Pine, to be covered with extra heavy **FIRE-CHIEF SHINGLE ROLL ROOFING, GUARANTEED BY US FOR FIFTEEN YEARS FOR WEAR AND COLOR.** Approved by National Board of Fire Underwriters. This gives you a dustproof, rainproof, snowproof and fire resisting roof. (See description of Fire-Chief Roofing on colored insert opposite page 1.)

"Double V Siding."

For a full description of our Doors, Windows, Hardware and Paint, refer to pages 43, 44, 47 and 54. A careful study of these details will convince you that in placing your order with us you are assured of getting your full money's worth. Remember, we are not trying to sell the lowest priced barn, but one of the highest quality of material and sound construction at the smallest outlay of money—quality considered. ONLY OUR GREAT RESOURCES AND OUR BUILDING EXPERIENCE ENABLES US TO QUOTE THE VERY LOW PRICES IN THIS CATALOG. REMEMBER, **THAT EVERY PIECE OF FRAMING LUMBER YOU RECEIVE FOR YOUR BARN IS "ALREADY CUT" AND FITTED AND THAT YOU GET COMPLETE PLANS SHOWING EXACTLY WHERE IT WILL FIT,** on receipt of order.

Optional Items.

On account of difference in requirements and arrangement of floor plans, we have **NOT INCLUDED** the following items in our catalog prices:

Ventilators and Cupolas.
Ventilating Systems.
Material for ground floors, because it may be desired to make them of concrete.
Inside Lining of the ground floor and ceiling.
Barn Equipment, such as stalls, partitions, bins, hay racks, hay carriers.
Special prices for such items will be quoted on application.

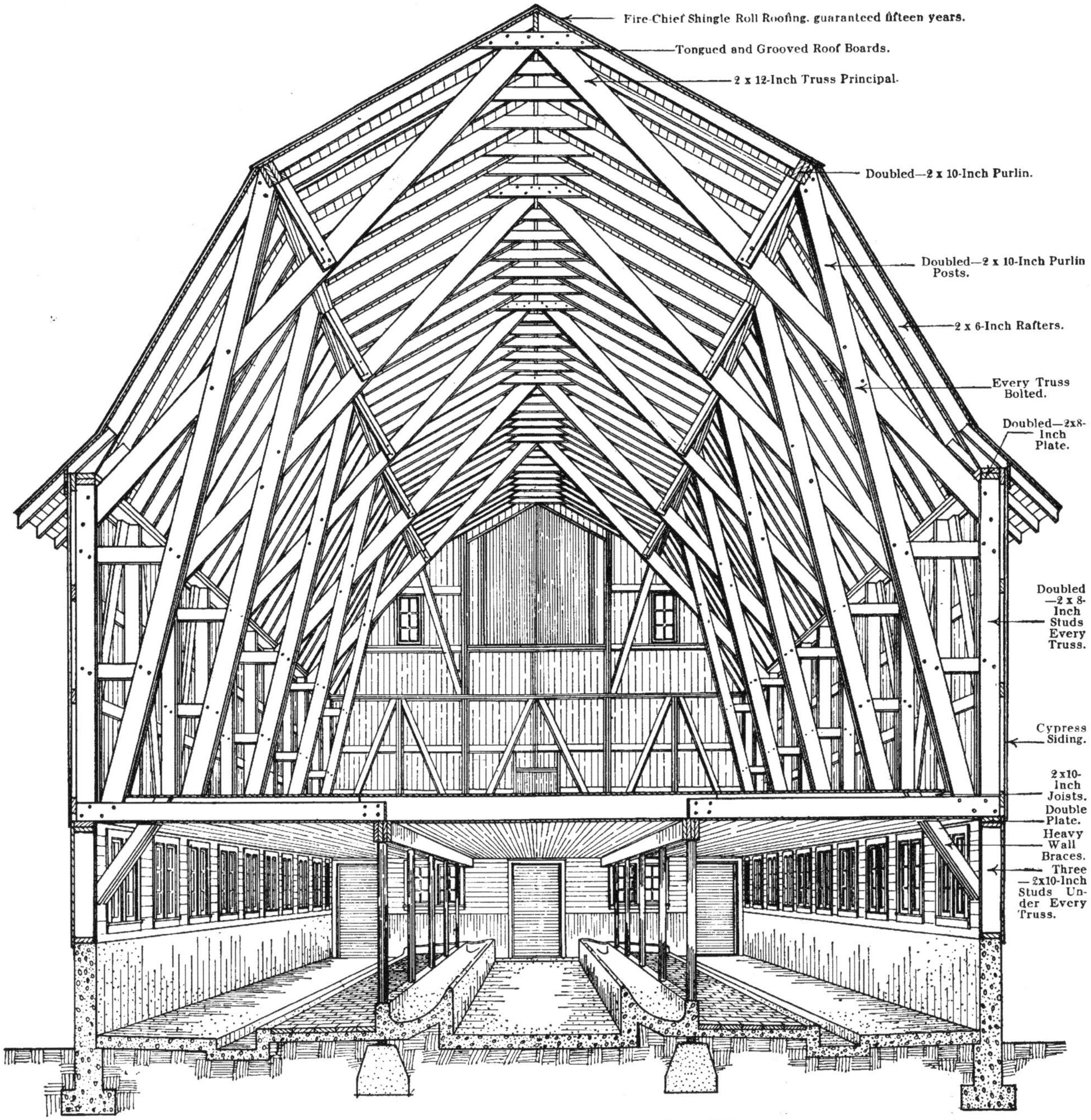

Trussed Roof Construction for Barn 36 Feet Wide.
Regarding equipment, inside lining, ceiling, etc., see "Optional Items" to the left.

Prices of the "Country Gentleman" Modern Barn No. 3007, "Already Cut" and Fitted.

Size of Barn: Width, Feet	Size of Barn: Length, Feet	Height From Sill to Eaves, Feet	Price of Barn: With No. 1 Common Cypress Siding ("Already Cut" and Fitted.)	Price of Barn: With Select Cypress Siding ("Already Cut" and Fitted.)	Haymow Capacity in Tons
32	32	8	$1,049.00	$1,080.00	32
32	38	8	1,176.00	1,208.00	38
32	44	8	1,302.00	1,337.00	44
32	50	8	1,429.00	1,464.00	50
32	56	8	1,555.00	1,593.00	56
32	62	8	1,682.00	1,721.00	62
32	68	8	1,808.00	1,850.00	68
34	38	8	1,225.00	1,262.00	42
34	44	8	1,355.00	1,394.00	48
34	50	8	1,486.00	1,527.00	55
34	56	8	1,616.00	1,660.00	61
34	62	8	1,747.00	1,791.00	69
34	68	8	1,877.00	1,923.00	75
34	74	8	2,008.00	2,057.00	81

Size of Barn: Width, Feet	Size of Barn: Length, Feet	Height From Sill to Eaves, Feet	Price of Barn: With No. 1 Common Cypress Siding ("Already Cut" and Fitted.)	Price of Barn: With Select Cypress Siding ("Already Cut" and Fitted.)	Haymow Capacity in Tons
34	80	8	$2,138.00	$2,188.00	88
36	38	8	1,372.00	1,413.00	45
36	44	8	1,441.00	1,484.00	51
36	50	8	1,582.00	1,624.00	58
36	56	8	1,721.00	1,766.00	65
36	62	8	1,860.00	1,907.00	72
36	68	8	1,999.00	2,049.00	79
36	74	8	2,139.00	2,190.00	86
36	80	8	2,279.00	2,332.00	93
36	86	8	2,418.00	2,473.00	100
36	92	8	2,559.00	2,613.00	107
36	98	8	2,697.00	2,755.00	114
36	104	8	2,836.00	2,896.00	121

ALREADY CUT AND FITTED PRICE $1,049.00 AND UP

An Imposing Modern Barn, Braced Rafter Construction, "Already Cut" and Fitted

THERE is a quiet charm about this modern farm building. It represents the approved type of American barn architecture, and it well deserves its name, the "Country Gentleman."

For the dairyman, for the horseman, or for mixed farming, this building is equally useful. It is designed to meet modern sanitary requirements and to permit of convenience in doing the daily work.

This barn is to be erected on a foundation of concrete or masonry extending 4 feet above the grade level. The ceiling height is 8 feet 6 inches.

Framing lumber is of No. 1 Yellow Pine.

The outside walls are sided with drop siding of Cypress, "The Wood Eternal." This secures a strong and weather resisting building.

To make the roof good and strong we furnish tight fitting tongued and grooved sheathing boards to be laid on the rafters and covered with Fire-Chief Shingle Roll Roofing, guaranteed by us for fifteen years. It resembles painted shingles.

Barns up to 44 feet long have one dormer on each side.

Barns 50 to 68 feet long have two dormers on each side.

Barns over 68 feet long have three dormers on each side.

Doors are made of Clear Cypress, "The Wood Eternal." Ground floor doors and the double hay door in the front gable slide on Roll Rite hangers, the best hangers we handle. A smaller hinged door on the mow floor level is convenient for many purposes. All doors are **"ready made,"** ready to hang in place, and are superior to those produced by hand carpentry.

Large windows, 1⅜ inches thick, with nine lights, are provided for the ground floor. Opening size of windows is 2 feet 8 inches by 3 feet 6 inches. Ventilation shields are furnished. A quick change of air can be made by lifting the windows back into the shields. The practical construction of the shields protects your stock from direct drafts.

Hardware, such as latches, bolts, screws, nails, etc., is included in the price, also sufficient paint for two coats, oxide red for the body and white for the trim. You may select other combinations from our paint page. (See page 47.) Ventilating systems, cupolas, barn equipment and foundation material are not included in the price.

Floor Plans. See pages 40 to 42 for sample floor plan arrangements. **Be sure to specify number of the floor plan that best suits your requirements.** Use the enclosed sketch sheet to outline your needs when changes are necessary and return for quotation on equipment.

Our "already cut" feature, together with **"ready made"** doors, sash, etc., furnished at the prices quoted, make it possible for you to erect this modern barn at a remarkable saving of time and labor. Free building plans, which are easy to understand, are furnished with every order.

You will appreciate the value, beauty, economy and convenience of this strong and sanitary modern barn.

ALREADY CUT AND FITTED PRICE $1,195.00 AND UP

A Modern Dairy Barn, Braced Rafter Type With Drop Siding.

NO dark corners in this Dairy Barn. Many long windows provide an abundance of light. Right here the first requirement of sanitation in a dairy barn has been carefully considered. The windows are equipped with ventilation shields; a quick change of air may be made by lifting windows back into these shields. No direct draft touches your stock. The arrangement of the ventilation shields protects them.

Sanitary steel columns with base and cap, for the support of the main girders, are furnished. These are the best columns for dairy barns.

This barn is constructed in the Braced Rafter Type (for full details see page 4). Height of ground floor, 8 feet 6 inches.

The framing lumber for this barn is No. 1 Yellow Pine.

The outside walls are covered with drop siding of Cypress, "The Wood Eternal." This combination secures a very strong frame and outside walls that will last a lifetime.

For the roof of the barn we furnish **tight fitting tongued and grooved sheathing boards** on which is to be laid **Fire-Chief Shingle Roll Roofing. Fire-Chief Roofing** is guaranteed by us for **fifteen years.**

It resembles painted shingles.

Barns up to 44 feet long have one dormer on each side.

Barns 50 to 68 feet long have two dormers on each side.

Barns over 68 feet long have three dormers on each side.

All doors in this barn are **"ready made"**—ready to hang in place and are superior to doors produced by hand carpentry. The doors on the ground floor are of double thickness. These doors and the hay door slide on Roll Rite Hangers, the highest priced hangers we handle. The smaller hinged door at the mow level will be found of great advantage.

Hardware, such as latches, bolts, nails, etc., is included in the above price. Sufficient **paint for two coats is furnished,** oxide red for the body and white for the trim. You may select any other combination from our paint page (see page 47). Foundation material, ventilating systems, cupolas, barn equipment and silos are not included in the price.

Our "already cut" feature, together with **"ready made"** doors, sash, etc., furnished at the prices quoted, enable you to build this barn at a big saving of time and labor. Free building plans, easy to understand, are furnished with each order.

Floor Plans. See pages 40 to 42 for sample floor plan arrangements. **Be sure to specify number of the floor plan that best suits your requirements.** Use the enclosed sketch sheet to outline your needs when changes are necessary and return for quotation on equipment.

Be sure to give size when ordering.

QUALITY GUARANTEED

Prices of the "Pride of the Farmstead" Modern Dairy Barn No. 2054 "Already Cut" and Fitted.

SIZE OF BARN			PRICE OF BARN			SIZE OF BARN			PRICE OF BARN		
Width, Feet	Length, Feet	Height From Sill to Eaves, Feet	With No. 1 Common Cypress Siding ("Already Cut" and Fitted.)	With Select Cypress Siding ("Already Cut" and Fitted.)	Haymow Capacity in Tons	Width, Feet	Length, Feet	Height From Sill to Eaves, Feet	With No. 1 Common Cypress Siding ("Already Cut" and Fitted.)	With Select Cypress Siding ("Already Cut" and Fitted.)	Haymow Capacity in Tons
30	32	12	$1,195.00	$1,247.00	31	34	74	16	$2,462.00	$2,565.00	106
30	38	12	1,345.00	1,410.00	37	34	86	16	2,794.00	2,901.00	123
30	44	12	1,493.00	1,560.00	43	34	98	16	3,119.00	3,236.00	140
30	50	12	1,643.00	1,706.00	49	34	110	16	3,443.00	3,571.00	157
32	32	14	1,263.00	1,325.00	37	36	38	16	1,548.00	1,622.00	61
32	38	14	1,420.00	1,485.00	44	36	44	16	1,717.00	1,796.00	70
32	44	14	1,575.00	1,645.00	52	36	50	16	1,887.00	1,970.00	80
32	50	14	1,730.00	1,770.00	59	36	56	16	2,055.00	2,144.00	90
32	56	14	1,885.00	1,966.00	66	36	62	16	2,225.00	2,306.00	99
32	62	14	2,041.00	2,126.00	73	36	74	16	2,563.00	2,667.00	118
34	38	16	1,490.00	1,559.00	54	36	86	16	2,901.00	3,016.00	138
34	44	16	1,654.00	1,727.00	63	36	98	16	3,239.00	3,364.00	157
34	50	16	1,816.00	1,894.00	71	36	110	16	3,577.00	3,713.00	176
34	56	16	1,980.00	2,061.00	80	36	122	16	3,915.00	4,062.00	195
34	62	16	2,142.00	2,229.00	89	36	146	16	4,592.00	4,759.00	234

Prices of the "Blue Ribbon" Modern Dairy Barn No. 2055, "Already Cut" and Fitted.

SIZE OF BARN			PRICE OF BARN		
Width, Feet	Length, Feet	Height, From Sill to Eaves, Feet	With No. 1 Common Cypress Siding	With Select Cypress Siding	Haymow Capacity in Tons
30	42	12	$1,430.00	$1,475.00	38
30	56	12	1,742.00	1,800.00	50
30	70	12	2,053.00	2,120.00	84
32	42	14	1,521.00	1,585.00	50
32	56	14	1,860.00	1,934.00	67
32	70	14	2,199.00	2,282.00	84
32	84	14	2,539.00	2,629.00	100
32	98	14	2,878.00	2,978.00	117
34	42	16	1,646.00	1,728.00	57
34	56	16	1,972.00	2,108.00	76
34	70	16	2,374.00	2,490.00	95
34	84	16	2,757.00	2,875.00	114
34	98	16	$3,127.00	$3,253.00	132
34	112	16	3,498.00	3,633.00	151
34	126	16	3,868.00	4,014.00	170
34	140	16	4,238.00	4,396.00	189
36	42	16	1,696.00	1,777.00	63
36	56	16	2,076.00	2,168.00	84
36	70	16	2,456.00	2,560.00	105
36	84	16	2,836.00	2,951.00	126
36	98	16	3,217.00	3,342.00	147
36	112	16	3,724.00	3,733.00	168
36	126	16	3,978.00	4,124.00	189
36	140	16	4,358.00	4,517.00	210

"Already Cut" and Fitted.

ALREADY CUT AND FITTED PRICE $1,430.00 AND UP

A Modern Dairy Barn of Trussed Roof Construction, With Vertical Siding.

THIS Modern Dairy Barn is designed to meet the requirements of those who prefer the Trussed Roof Construction. (See page 5.) This barn has an abundance of long windows which provide plenty of light and air for the ground floor. The windows are equipped with ventilation shields. A quick change of air can be made by lifting the windows back into the shields. No direct draft touches your stock. The arrangement of the ventilation shields protects them. Care has been taken to work out these essentials of light and air to obtain the most sanitary conditions for this Dairy Barn. Our arrangement meets all requirements. Height of ground floor, 8 feet 6 inches.

Sanitary Steel Columns with base and cap for the support of the main girders, are also provided.

The framing lumber for this barn is No. 1 Yellow Pine. All trusses are to be bolted together. For the outside walls vertical siding of the double V pattern of Cypress, "The Wood Eternal," is furnished. This combination secures a very strong frame and outside walls that will withstand the weather for a lifetime.

Realizing that the roof of a barn being hard to get at is, therefore, the most neglected, we furnish **tight fitting, tongued and grooved sheathing boards** to cover the rafters and **Fire-Chief Shingle Roll Roofing. Fire-Chief Roofing** is guaranteed by us for **fifteen years.** It resembles painted shingles.

Barns up to 42 feet long have one dormer on each side.

Barns 56 to 70 feet long have two dormers on each side.

Barns over 70 feet long have three dormers on each side.

The doors are **"ready made"**—ready to hang in place and are superior to those produced by hand carpentry. The doors on the first floor are of **double** thickness. The doors on the first floor, also the hay door in the front gable, slide on Roll Rite Hangers, the highest priced hangers we handle. Consider the advantage of the little hinged door at the mow level.

Hardware, such as latches, bolts, nails, etc., is included in the above price. Sufficient **paint for two coats** is furnished, oxide red for the body and white for the trim. You may select any other combination from our paint page. (See page 47.) Ventilating systems, cupolas, barn equipment, silos, etc., are not included in the price.

Our "already cut" feature, together with factory **"ready made"** doors, sash, etc., furnished at the prices quoted, make it possible for you to build this high class Dairy Barn at a remarkable saving of time and labor.

Floor Plans. See pages 40 to 42 for sample floor plan arrangements. **Be sure to specify number of the floor plan that best suits your requirements.** Use the enclosed sketch sheet to outline your needs when changes are necessary and return for quotation on equipment.

ALREADY CUT AND FITTED PRICE $705.00 AND UP

A Modern Barn With Gothic Roof Construction "Already Cut" and Fitted.

OUR "Cyclone" Barn combines good appearance with strong construction and ample storage capacity. It offers great resistance to wind pressure and is designed principally for such sections of the country where storms and high winds are frequent. It may be used with equal advantage as a dairy barn, a horse barn or a barn for mixed stock. The barn should be erected on a 2-foot foundation of concrete or masonry. The height of the ground floor is 8 feet 6 inches. It has a solid and spacious mow floor above, made of tongued and grooved No. 1 Yellow Pine flooring, which will also keep the stable below free from dust and drafts.

All framing lumber is of No. 1 Yellow Pine.

The outside walls are covered with double V pattern vertical siding of Cypress, "The Wood Eternal." This combination secures a very strong frame and outside walls that will withstand the weather for a lifetime.

The roof arches are good and strong, easy to build up, and provide an unobstructed mow. The arches are to be covered with tight fitting tongued and grooved sheathing boards which we furnish, over which is to be laid **Fire-Chief Shingle Roll Roofing. Fire-Chief Roofing** is guaranteed by us for **fifteen years.** It resembles painted shingles.

Doors are made of Clear Cypress, "The Wood Eternal." Ground floor doors slide on Roll Rite Hangers, the highest priced hangers we handle. The haymow door has three extra heavy strap hinges. A smaller door on the mow floor level is convenient for many purposes. All doors are **"ready made"**—ready to hang in place and are superior to those produced by hand carpentry.

Large windows, 1⅜ inches thick, with nine lights, are liberally provided for the ground floor. Opening size of windows is 32x42 inches. Ventilation shields are furnished. A quick change of air can be made by lifting the windows back into the shields. The practical construction of the shields protects your stock from direct drafts.

For barns 24 feet wide we furnish a single door at the corners instead of the driveway doors in the center.

Hardware, such as latches, bolts, screws, nails, etc., is included in the price, also sufficient **paint for two coats,** oxide red for the body and white for the trim. You may select other combinations from our paint page (see page 47). Ventilating systems, cupolas, barn equipment, silos, foundation material, etc., are not included in the price.

Floor Plans. See pages 40 to 42 for sample floor plan arrangements. **Be sure to specify number of the floor plan that best suits your requirements.** Use the enclosed sketch sheet to outline your needs when changes are necessary and return for quotation on equipment.

Be sure to notice that all our framing lumber is "already cut" and fitted, that our doors are all **made and ready to hang,** and that our roofing is guaranteed for fifteen years. **Take no substitute for Cypress Siding.**

QUALITY GUARANTEED

Prices of the "Cyclone" Modern Barn No. 3026, "Already Cut" and Fitted.

Size of Barn: Width, Feet	Size of Barn: Length, Feet	Size of Barn: Height From Sill to Eaves, Feet	Price of Barn: With No. 1 Common Cypress Siding	Price of Barn: With Select Cypress Siding	Haymow Capacity in Tons, About
24	24	8	$ 705.00	$ 732.00	14
24	32	8	878.00	907.00	16
24	38	8	978.00	1,009.00	19
24	44	8	1.077.00	1,111.00	22
24	50	8	1,176.00	1,214.00	25
28	32	8	1,022.00	1,058.00	22
28	38	8	1,139.00	1,177.00	26
28	44	8	1,261.00	1,298.00	30
28	50	8	1,376.00	1,416.00	34
28	56	8	1,495.00	1,537.00	38
30	32	8	1,085.00	1,125.00	25
30	38	8	1,213.00	1,291.00	30
30	44	8	1,339.00	1,420.00	35
30	50	8	1,467.00	1,548.00	40
30	56	8	1,594.00	1,677.00	45
30	62	8	1,666.00	1,806.00	50
32	32	8	1.162.00	1,205.00	27
32	38	8	1,301.00	1,347.00	32
32	44	8	1,440.00	1,489.00	37
32	50	8	1,579.00	1,630.00	42
32	56	8	1,720.00	1,774.00	47
32	62	8	1,859.00	1,915.00	52
32	68	8	1,998.00	2,057.00	57
32	80	8	2,125.00	2,199.00	62
34	38	8	1,420.00	1,467.00	42
34	44	8	1,573.00	1,623.00	50
34	50	8	1,725.00	1,776.00	57
34	56	8	1,878.00	1,930.00	64
34	62	8	2,030.00	2,083.00	71

"Already Cut" and Fitted. "Already Cut" and Fitted.

Size of Barn: Width, Feet	Size of Barn: Length, Feet	Size of Barn: Height From Sill to Eaves, Feet	Price of Barn: With No. 1 Common Cypress Siding	Price of Barn: With Select Cypress Siding	Haymow Capacity in Tons, About
34	68	8	$2,183.00	$2,237.00	78
34	74	8	2,336.00	2,392.00	85
34	86	8	2,641.00	2,700.00	98
34	98	8	2,947.00	3,008.00	110
34	110	8	3,253.00	3,317.00	123
36	38	8	1,470.00	1,520.00	50
36	44	8	1,628.00	1,681.00	57
36	50	8	1,786.00	1,842.00	65
36	56	8	1,944.00	2,002.00	72
36	62	8	2,090.00	2,162.00	79
36	68	8	2,248.00	2,323.00	86
36	80	8	2,563.00	2,647.00	100
36	92	8	2,876.00	2,967.00	114
36	104	8	3,195.00	3,288.00	128
36	116	8	3.511.00	3,607.00	142
36	128	8	3,828.00	3,931.00	156
36	140	8	4,144.00	4,253.00	170
40	44	8	1,820.00	1,888.00	56
40	50	8	1,991.00	2,058.00	64
40	56	8	2,161.00	2,230.00	70
40	62	8	2,332.00	2,400.00	76
40	68	8	2,500.00	2,573.00	83
40	74	8	2,672.00	2,743.00	90
40	80	8	2,841.00	2,886.00	96
40	92	8	3,181.00	3,258.00	110
40	104	8	3,522.00	3,601.00	124
40	116	8	3,761.00	3,942.00	138
40	128	8	4,200.00	4,285.00	152
40	140	8	4,540.00	4,628.00	167

"Already Cut" and Fitted. "Already Cut" and Fitted.

ALREADY CUT AND FITTED PRICE $1,587.00 AND UP

A Modern Dairy Barn, With Extra Large Haymow Capacity, Trussed Roof Construction, "Already Cut" and Fitted.

THIS barn is designed to meet the requirements of a dairy barn with extra large mow capacity. We have raised the concrete foundation four feet above the established grade level. On the foundation a sill is bolted, and on this the framing lumber for the windows and double plate for the support of the upper structure are erected. The framing is exactly as shown in the illustration on page 5. Ground floor height, 8 feet 6 inches. The trusses are held together by strong bolts, which we furnish. The capacity is very large and sufficient to meet more than ordinary requirements.

The whole construction is ingeniously conceived and carried through with simplicity and skill. The barn is of very imposing appearance on account of its great dimensions.

Long windows provide plenty of light and air for the ground floor. The windows are equipped with ventilation shields. Air can be quickly changed by lifting the windows back into the shields. The arrangement of the ventilation shields protects your stock from direct drafts. **Sanitary Steel Columns** with base and cap support the main girders.

All framing lumber is No. 1 Yellow Pine.

For the outside walls we furnish Cypress drop siding up to the mow floor, and Cypress double V vertical siding for the upper part. This combination secures a very strong frame and outside walls that will last a lifetime.

For the roof of the barn we furnish **tight fitting tongued and grooved sheathing boards** on which is to be laid **Fire-Chief Shingle Roll Roofing. Fire-Chief Roofing** is guaranteed for **fifteen years.** It resembles painted shingles.

Barns 42 feet long have one dormer on each side.

Barns 56 to 70 feet long have two dormers on each side.

Barns over 70 feet long have three dormers on each side.

Doors are made of Clear Cypress. Ground floor doors are double thickness. Ground floor doors and the hay door slide on Roll Rite Hangers, the highest priced hangers we handle. A little hinged door at the mow level in one gable will be found of great convenience. All doors are **"ready made"**—ready to hang in place, and are superior to doors produced by hand carpentry.

We furnish all the necessary hardware and sufficient **paint for two coats,** oxide red for the body and white for the trim. Other combinations may be selected from our paint page (see page 47). Ventilating systems, cupolas, equipment, concrete silos, etc., are not included in price.

Floor Plans. See pages 40 to 42 for sample floor plan arrangements. **Be sure to specify number of the floor plan that best suits your requirements.** Use the enclosed sketch sheet to outline your needs when changes are necessary and return for quotation on equipment.

The "already cut" feature of this barn, together with **"ready made"** doors, sash, etc., at the price quoted, make it possible for you to erect this high class Dairy Barn at the lowest cost and at a big saving of time and labor. Building plans, which are easy to understand, are furnished with every order.

Prices of "Dairy Belle" Modern Dairy Barn No. 2057, "Already Cut" and Fitted.

SIZE OF BARN			PRICE OF BARN ("Already Cut" and Fitted.)		
Width, Feet	Length, Feet	Height From Sill to Eaves, Feet	With No. 1 Common Cypress Siding	With Select Siding	Haymow Capacity in Tons
30	42	18	$1,587.00	$1,667.00	61
30	56	18	1,948.00	2,042.00	81
30	70	18	2,311.00	2,417.00	101
32	42	18	1,678.00	1,766.00	71
32	56	18	2,058.00	2,158.00	95
32	70	18	2,437.00	2,550.00	118
32	84	18	2,817.00	2,942.00	142
32	98	18	3,196.00	3,334.00	166
34	42	18	1,787.00	1,878.00	71
34	56	18	2,191.00	2,295.00	96
34	70	18	2,595.00	2,712.00	120
34	84	18	2,999.00	3,128.00	144
34	98	18	$3.403.00	$3,545.00	168
34	112	18	3,807.00	3,961.00	192
34	126	18	4,211.00	4,378.00	215
34	140	18	4,614.00	4,795.00	240
36	42	18	1,854.00	1,948.00	77
36	56	18	2,279.00	2,386.00	102
36	70	18	2,703.00	2,823.00	127
36	84	18	3,127.00	3,260.00	153
36	98	18	3,552.00	3,697.00	178
36	112	18	3,976.00	4,134.00	204
36	126	18	4,400.00	4,571.00	229
36	140	18	4.825.00	5,009.00	255

ALREADY CUT AND FITTED PRICE $1,707.00 AND UP

A Modern Dairy Bridge Barn, With Large Storage Capacity, Truss Construction, "Already Cut" and Fitted.

THE attempt to construct a Bridge or Bank Dairy Barn meeting modern requirements of sanitation has long been a problem. Most bank barns are impractical as dairy barns because the ground floor is partially surrounded by earth walls. Such arrangement bars a great part of light and air from the ground floor. Without plenty of light and air a modern dairy barn is an impossibility. The great demand for a dairy barn combining access to the mow with team and load, a very large storage capacity and modern sanitary ground floor arrangements has prompted us to design this barn. Ground floor height, 8 feet 6 inches.

The barn is not to be built close against a bank. Sufficient space for light and air has to be left between the bank and the side wall of the barn. The intervening space is to be overbridged where the large driveway doors are located. (The bridge is not included in the price.)

Plenty of long windows on **every side** of the barn provide an abundance of light. The windows are equipped with ventilation shields. A quick change of air can be made by lifting the windows back into the ventilation shields. No direct draft touches the stock. The ventilation shields protect them. This arrangement solves the problem of a modern dairy bridge or bank barn satisfactorily.

Sanitary Tubular Steel Columns for the support of the main girders are also furnished.

The foundation walls are raised 4 feet above the grade level.

The framing lumber is No. 1 Yellow Pine. All trusses are to be bolted together.

For the outside walls we furnish Cypress drop siding up to the mow floor and Cypress double V vertical siding for the upper part. This combination secures a very strong frame and outside walls that will last a lifetime.

For the roof of the barn we furnish **tight fitting tongued and grooved sheathing boards to be laid on the rafters and covered with Fire-Chief Shingle Roll Roofing. Fire-Chief Roofing** is guaranteed by us for **fifteen years.** It resembles painted shingles.

Barns 42 feet long have one dormer on each side.

Barns 56 and 70 feet long have two dormers on each side.

Barns over 70 feet long have three dormers on each side.

Doors are made of Clear Cypress. The ground floor doors are of double thickness and slide on Roll Rite Hangers, the highest priced hangers we handle. The driveway doors also slide on Roll Rite Hangers and are large enough for big loads to enter the mow. The small built in door in one of the driveway doors is convenient for personal entrance to the mow without the necessity of opening the large doors (see page 43). The advantage of the hinged door at the mow level is plain. All doors are **"ready made"**—ready to hang in place and are superior to those produced by hand carpentry.

In addition, we furnish all the necessary hardware, such as latches, bolts, screws, nails, etc., and sufficient **paint for two coats,** oxide red for the body and white for the trim. Other combinations may be selected from our paint page (see page 47). Ventilating systems, cupolas, equipment, silos, concrete, etc., are not included in the price.

Floor Plans. See pages 40 to 42 for sample floor plan arrangements. **Be sure to specify number of the floor plan that best suits your requirements.** Use the enclosed sketch sheet to outline your needs when changes are necessary and return for quotation on equipment.

Free building plans, which are easy to understand, are furnished with every order.

QUALITY GUARANTEED

Our "already cut" feature, together with factory **"ready made"** doors, sash, etc., enables you to build this modern bridge or bank dairy barn at a big saving of time and labor. Be sure to give size of barn when ordering.

Prices of "Dairy Queen" Modern Dairy Barn No. 2058, "Already Cut" and Fitted.

SIZE OF BARN			PRICE OF BARN		HAYMOW	SIZE OF BARN			PRICE OF BARN		HAYMOW
Width, Feet	Length, Feet	Height From Sill to Eaves, Feet	With No. 1 Common Cypress Siding	With Select Cypress Siding	Capacity in Tons	Width, Feet	Length, Feet	Height From Sill to Eaves, Feet	With No. 1 Common Cypress Siding	With Select Cypress Siding	Capacity in Tons
30	42	18	$1,707.00	$1,771.00	61	34	98	18	$3,472.00	$3,602.00	168
30	56	18	2,068.00	2,145.00	81	34	112	18	3,876.00	4,020.00	192
30	70	18	2,429.00	2,519.00	101	34	126	18	4,280.00	4,436.00	215
32	42	18	1,785.00	1,866.00	71	34	140	18	4,683.00	4,852.00	240
32	56	18	2,164.00	2,257.00	95	36	42	18	1,966.00	2,055.00	77
32	70	18	2,541.00	2,648.00	118	36	56	18	2,388.00	2,491.00	102
32	84	18	2,920.00	3,039.00	142	36	70	18	2,810.00	2,926.00	127
32	98	18	3,298.00	3,430.00	166	36	84	18	3,231.00	3,361.00	153
34	42	18	1,856.00	1,936.00	72	36	98	18	3,603.00	3,795.00	178
34	56	18	2,263.00	2,352.00	96	36	112	18	4,075.00	4,231.00	204
34	70	18	2,652.00	2,770.00	120	36	126	18	4,497.00	4,666.00	229
34	84	18	3,069.00	3,186.00	144	36	140	18	4,919.00	5,101.00	255

"Already Cut" and Fitted.

NOTE— Special door opposite driveway doors on mow floor, opening size 6 feet by 12 feet, on Roll Rite Hangers and Track, with barn door latch, complete, **$17.00** extra.

ALREADY CUT AND FITTED PRICE $841.00 AND UP

A Low Priced, Modern Barn, Trussed Roof Construction, "Already Cut" and Fitted

THIS barn is a bargain. It is designed for a flexible and convenient arrangement, providing solid comfort and sanitary conditions for a mixed stock.

It is of the trussed roof type as shown on page 5. The trusses are to be bolted together to secure greatest strength. We furnish the necessary bolts. Ground floor height, 8 feet 6 inches.

All framing lumber is of No. 1 yellow pine.

The outside walls are covered with double V vertical siding of Cypress, "The Wood Eternal." This combination secures a very strong frame and outside walls that will last a lifetime.

To make the roof good and strong, we furnish **tight fitting, tongued and grooved sheathing boards** to be laid on the rafters and covered with **Fire-Chief Shingle Roll Roofing. Fire-Chief Roofing** is guaranteed by us for **fifteen years.** It resembles painted shingles.

Doors are made of Clear Cypress, "The Wood Eternal." Ground floor doors and the double hay door slide on Roll Rite Hangers, the highest priced hangers we handle. A smaller hinged door on the mow floor level is convenient for many purposes. All doors are **"ready made"**—ready to hang in place and are superior to those produced by hand carpentry.

Large windows, 1⅜ inches thick, with nine lights, are liberally provided for the ground floor. Opening size of windows, 2 feet 8 inches by 3 feet 6 inches. Ventilation shields are furnished. A quick change of air can be made by lifting the windows back into the shields. The practical construction of the shields protects your stock from direct drafts.

End arrangement of barns, 24 feet wide, is shown in the small illustration. This size has a slightly different door and window arrangement on ground floor and a drop Hay Door.

Hardware, such as latches, bolts, screws, nails, etc., is included in the price, also sufficient **paint for two coats,** oxide red for the body and white for the trim. You may select other combinations from our paint page (see page 47). Ventilating systems, cupolas, barn equipment, silos, foundation material, etc., are not included in the price.

Floor Plans. See pages 40 to 42 for sample floor plan arrangements. **Be sure to specify number of the floor plan that best suits your requirements.** Use the enclosed sketch sheet to outline your needs when changes are necessary and return for quotation on equipment.

Our "already cut" feature, together with **"ready made"** doors, sash, etc., furnished at the prices quoted, make it possible for you to erect this high class general purpose barn at a remarkable saving of time and labor. Free building plans, which are easy to understand, are furnished with every order.

Order now and save money.
Be sure to give size.

QUALITY GUARANTEED

Prices of "Favorite" Modern General Purpose Barn No. 2059, "Already Cut" and Fitted.

SIZE OF BARN			PRICE OF BARN		HAY MOW Capacity in Tons	SIZE OF BARN			PRICE OF BARN		HAY MOW Capacity in Tons
Width, Feet	Length, Feet	Height From Sill to Eaves, Feet	With No. 1 Common Cypress Siding	With Select Cypress Siding		Width, Feet	Length, Feet	Height From Sill to Eaves, Feet	With No. 1 Common Cypress Siding	With Select Cypress Siding	
24	28	12	$ 841.00	$ 878.00	17	34	98	16	$2,806.00	$2,930.00	132
24	42	12	1,085.00	1,130.00	25	34	112	16	3,131.00	3,264.00	151
24	56	12	1,330.00	1,383.00	34	34	126	16	3,454.00	3,600.00	170
28	28	12	954.00	997.00	24	34	140	16	3,778.00	3,935.00	189
28	42	12	1,222.00	1,272.00	36	36	42	16	1,579.00	1,662.00	76
28	56	12	1,490.00	1,548.00	47	36	56	16	1,911.00	2,005.00	101
28	70	12	1,757.00	1,823.00	59	36	70	16	2,242.00	2,348.00	126
30	42	12	1,287.00	1,346.00	38	36	84	16	2,567.00	2,690.00	151
30	56	12	1,562.00	1,628.00	50	36	98	16	2,905.00	3,033.00	176
30	70	12	1,836.00	1,911.00	63	36	112	16	3,237.00	3,376.00	202
30	84	12	2,109.00	2,192.00	76	36	126	16	3,569.00	3,720.00	227
32	42	14	1,431.00	1,509.00	51	36	140	16	3,900.00	4,062.00	252
32	56	14	1,728.00	1,815.00	67	40	42	16	1,733.00	1,825.00	74
32	70	14	2,024.00	2,122.00	84	40	56	16	2,095.00	2,198.00	99
32	84	14	2,321.00	2,428.00	101	40	70	16	2,454.00	2,571.00	123
32	98	14	2,618.00	2,734.00	118	40	84	16	2,816.00	2,942.00	149
34	42	16	1,512.00	1,587.00	57	40	98	16	3,177.00	3,315.00	174
34	56	16	1,835.00	1,924.00	76	40	112	16	3,538.00	3,686.00	198
34	70	16	2,159.00	2,259.00	95	40	126	16	3,898.00	4,059.00	223
34	84	16	2,483.00	2,582.00	113	40	140	16	4,259.00	4,431.00	248

ALREADY CUT AND FITTED PRICE $757.00 AND UP

A Modern General Purpose Barn, Braced Rafter Type, With Drop Siding.

LIGHT, air and comfort for a mixed stock are the special features of this General Purpose Barn. The construction is of the braced rafter type shown on page 4.

Ground floor height, 8 feet 6 inches.

Framing lumber is of No. 1 yellow pine.

The outside walls are covered with drop siding of Cypress, "The Wood Eternal." This secures a strong and weather resisting combination.

To make the roof good and strong we furnish **tight fitting tongued and grooved sheathing boards** to be laid on the rafters and covered with **Fire-Chief Shingle Roll Roofing. Fire-Chief Roofing** is guaranteed by us for **fifteen years.** It resembles painted shingles.

Doors are made of Clear Cypress, "The Wood Eternal." Ground floor doors and the double hay door slide on Roll Rite Hangers, the highest priced hangers we handle. A smaller hinged door on the mow floor level is convenient for many purposes. All doors are **"ready made"**—ready to hang in place and are superior to those produced by hand carpentry.

Large windows, 1⅜ inches thick, with nine lights, are liberally provided for the ground floor. Opening size of windows is 2 feet 8 inches by 3 feet 6 inches. Ventilation shields are furnished. A quick change of air can be made by lifting the windows back into the shields. The practical construction of the shields protects your stock from direct drafts.

End arrangement of barn, 24 feet wide, is shown in the small illustration. This size has different door and window arrangement on ground floor and a drop hay door.

Hardware, such as latches, bolts, screws, nails, etc., is included in the price, also sufficient **paint for two coats,** oxide red for the body and white for the trim. You may select other combinations from our paint page (see page 47). Ventilating systems, cupolas, barn equipment, silos, foundation material, etc., are not included in the price.

Floor Plans. See pages 40 to 42 for sample floor plan arrangements. **Be sure to specify number of the floor plan that best suits your requirements.** Use the enclosed sketch sheet to outline your needs when changes are necessary and return for quotation on equipment.

Our "already cut" feature, together with **"ready made"** doors, sash, etc., furnished at the prices quoted, make it possible for you to erect this modern general purpose barn at a remarkable saving of time and labor. Free building plans, which are easy to understand, are furnished with every order.

QUALITY GUARANTEED

You will appreciate the value, economy and convenience in this strong and sanitary modern general purpose barn.

Prices of "Ideal" Modern General Purpose Barn No. 2060, "Already Cut" and Fitted.

Size of Barn: Width, Feet	Size of Barn: Lgth., Feet	Size of Barn: Height, From Sill to Eaves, Feet	Price of Barn: With No. 1 Common Cypress Siding ("Already Cut" and Fitted.)	Price of Barn: With Select Cypress Siding ("Already Cut" and Fitted.)	Hay-Mow Capacity in Tons
24	26	12	$ 757.00	$ 791 00	17
24	32	12	856.00	895.00	21
24	38	12	956.00	999.00	25
24	44	12	1,055.00	1,097.00	29
28	32	12	1,025.00	1,071.00	29
28	38	12	1,142.00	1,192.00	34
28	44	12	1,260.00	1,313.00	39
28	56	12	1,493.00	1,555.00	49
30	32	12	1,120.00	1,171.00	29
30	38	12	1,245.00	1,301.00	34
30	44	12	1,370.00	1,430.00	43
30	50	12	1,494.00	1,559 00	49
30	62	12	1,758.00	1,829.00	61
30	74	12	1,994.00	2,074.00	72
32	32	14	1,170.00	1,219.00	38
32	38	14	1,290.00	1,353.00	45
32	44	14	1,418.00	1,494.00	51
32	56	14	1,676.00	1,763.00	66
32	62	14	1,805.00	1,884.00	73
32	74	14	2,061.00	2,150.00	87
34	38	16	1,379.00	1,448 00	55

Size of Barn: Width, Feet	Size of Barn: Lgth., Feet	Size of Barn: Height, From Sill to Eaves, Feet	Price of Barn: With No. 1 Common Cypress Siding ("Already Cut" and Fitted.)	Price of Barn: With Select Cypress Siding ("Already Cut" and Fitted.)	Hay Mow Capacity in Tons
34	44	16	$1,521.00	$1,594.00	63
34	56	16	1,804.00	1,888.00	80
34	68	16	2,086.00	2,180.00	97
34	74	16	2,228.00	2,325.00	106
34	86	16	2,511.00	2,619.00	123
34	98	16	2,794.00	2,911 00	140
36	38	16	1,437.00	1,510.00	61
36	44	16	1,584.00	1,662.00	71
36	56	16	1,876.00	1,965.00	90
36	68	16	2,169.00	2,267.00	109
36	74	16	2,315.00	2,419.00	118
36	86	16	2,609.00	2,721.00	138
36	98	16	2,902.00	3,025.00	157
36	110	16	3,138.00	3,327.00	176
36	146	16	4,074.00	4,236.00	234
40	56	16	2,067.00	2,161.00	104
40	68	16	2,381.00	2,487.00	125
40	98	16	3,170 00	3,302.00	180
40	110	16	3,486.00	3,629.00	202
40	134	16	4,116.00	4,281.00	247
40	152	16	4,590.00	4,769.00	280

ALREADY CUT AND FITTED PRICE $804.00 AND UP

A General Purpose Barn, With Gothic Roof Construction and Vertical V Siding.

THIS barn is as strong as it is attractive (see front cover). It is easy to build and requires no expensive labor. The haymow is large and unobstructed. Simple directions which go with every barn will fully explain the manner of erecting it.

Wherever a Gothic Roof Barn is erected, the builder is very proud of it. His neighbors want one like it. Our customers who put up this barn will have a building that will be admired by everybody. Height of ground floor, 8 feet 6 inches.

All framing lumber is of No. 1 Yellow Pine.

The outside walls are covered with double V pattern vertical siding of cypress, "The Wood Eternal." This combination secures a very strong frame and outside walls that will withstand the weather for a lifetime.

The roof arches are good and strong, easy to build up, and provide an unobstructed mow. The arches are to be covered with tight fitting tongued and grooved sheathing boards which we furnish, over which is to be laid **Fire-Chief Shingle Roll Roofing. Fire-Chief Roofing** is guaranteed by **us** for **fifteen years.** It resembles painted shingles.

Doors are made of Clear Cypress, "The Wood Eternal." Ground floor doors slide on Roll Rite Hangers, the highest priced hangers we handle. The haymow door has three extra heavy strap hinges. A smaller door on the mow floor level is convenient for many purposes. All doors are **"ready made"**—ready to hang in place and are superior to those produced by hand carpentry.

Large windows, 1⅜ inches thick, with nine lights, are liberally provided for the ground floor. Opening size of windows is 2 feet 8 inches by 3 feet 6 inches. Ventilation shields are furnished. A quick change of air can be made by lifting the windows back into the shields. This protects your stock from direct drafts.

Different door arrangements are furnished at the ends of barns 24 feet wide (see small illustration).

Hardware, such as latches, bolts, screws, nails, etc., is included in the price, also sufficient **paint for two coats,** oxide red for the body and white for the trim. You may select other combinations from our paint page (see page 47). Ventilating systems, cupolas, barn equipment, silos, foundation material, etc., are not included in the price.

Floor Plans. See pages 40 to 42 for sample floor plan arrangements. **Be sure to specify number of the floor plan that best suits your requirements.** Use the enclosed sketch sheet to outline your needs when changes are necessary and return for quotation on equipment.

Can be furnished with Cross Driveway with 12x12-foot doors, with no mow floor over Cross Driveway, for $18 extra.

Be sure to notice that all our framing timber is already cut and fitted, that our doors are all **made and ready to hang,** that our roofing is guaranteed for fifteen years. **Take no substitute for Cypress Siding. When ordering barn be sure to give size.**

QUALITY GUARANTEED

Prices of "Our Leader" Modern General Purpose Barn No. 2061, "Already Cut" and Fitted.

Size of Barn: Width, Feet	Size of Barn: Length, Feet	Size of Barn: Height From Sill to Eaves, Feet	Price of Barn: With No. 1 Common Cypress Siding ("Already Cut" and Fitted.)	Price of Barn: With Select Cypress Siding ("Already Cut" and Fitted.)	HAYMOW Capacity in Tons	Size of Barn: Width, Feet	Size of Barn: Length, Feet	Size of Barn: Height From Sill to Eaves, Feet	Price of Barn: With No. 1 Common Cypress Siding ("Already Cut" and Fitted.)	Price of Barn: With Select Cypress Siding ("Already Cut" and Fitted.)	HAYMOW Capacity in Tons
24	24	12	$ 804.00	$ 841.00	18	34	68	16	$2.442.00	$2.545.00	111
24	32	12	948.00	989.00	22	34	74	16	2.610.00	2.718.00	121
24	38	12	1.057.00	1.101.00	26	34	86	16	2.946.00	3.063.00	140
24	44	12	1.165.00	1.213.00	30	34	98	16	3.282.00	3.408.00	160
24	50	12	1.273.00	1.324.00	35	34	110	16	3.618.00	3.754.00	180
28	32	12	1.099.00	1.147.00	30	36	38	16	1.656.00	1.743.00	69
28	38	12	1.224.00	1.276.00	35	36	44	16	1.830.00	1.921.00	79
28	44	12	1.349.00	1.406.00	41	36	50	16	2.004.00	2.099.00	90
28	50	12	1.475.00	1.535.00	47	36	56	16	2.177.00	2.277.00	100
28	56	12	1.601.00	1.663.00	52	36	62	16	2.351.00	2.444.00	112
30	32	12	1.163.00	1.218.00	34	36	68	16	2.513.00	2.622.00	123
30	38	12	1.299.00	1.356.00	40	36	80	16	2.873.00	2.993.00	144
30	44	12	1.433.00	1.495.00	46	36	92	16	3.220.00	3.349.00	166
30	50	12	1.568.00	1.633.00	52	36	104	16	3.580.00	3.707.00	187
30	56	12	1.702.00	1.771.00	58	36	116	16	3.915.00	4.063.00	209
30	62	12	1.838.00	1.909.00	64	36	128	16	4.262.00	4.358.00	230
32	32	14	1.284.00	1.347.00	43	36	140	16	4.610.00	4.779.00	252
32	38	14	1.434.00	1.502.00	51	40	44	16	2.034.00	2.137.00	88
32	44	14	1.586.00	1.659.00	59	40	50	16	2.219.00	2.327.00	100
32	50	14	1.738.00	1.814.00	67	40	56	16	2.403.00	2.517.00	112
32	56	14	1.890.00	1.969.00	75	40	62	16	2.588.00	2.705.00	124
32	62	14	2.041.00	2.124.00	83	40	68	16	2.773.00	2.896.00	136
32	68	14	2.192.00	2.280.00	91	40	74	16	2.958.00	3.101.00	148
32	80	14	2.495.00	2.591.00	107	40	80	16	3.142.00	3.276.00	160
34	38	16	1.602.00	1.682.00	62	40	92	16	3.513.00	3.655.00	184
34	44	16	1.770.00	1.854.00	72	40	104	16	3.882.00	4.035.00	208
34	50	16	1.938.00	2.027.00	82	40	116	16	4.251.00	4.414.00	232
34	56	16	2.106.00	2.200.00	91	40	128	16	4.621.00	4.792.00	256
34	62	16	2.274.00	2.373.00	101	40	140	16	4.990.00	5.172.00	280

ALREADY CUT AND FITTED PRICE $757.00 AND UP

An Inexpensive Modern General Purpose Barn, Braced Rafter Construction, Drop Siding.

HERE is a modern barn that is within the reach of all. While every farm building in this catalog is priced very low when quality is considered, this General Purpose Barn, "already cut" and fitted, is an exceptional money saving value. It is suitable for all sections of the country and will give extra good service on any farm.

This barn is of the braced rafter construction (see page 4). Ground floor height, 8 feet 6 inches.

All framing lumber is of No. 1 Yellow Pine.

The outside walls are covered with drop siding of Cypress, "The Wood Eternal." This combination secures a very strong frame and outside walls that will last a lifetime.

To make the roof good and strong, we furnish **tight fitting tongued and grooved sheathing boards** to be laid on the rafters and covered with **Fire-Chief Shingle Roll Roofing. Fire-Chief Roofing** is guaranteed by us for **fifteen years.** It resembles painted shingles.

Doors are made of Clear Cypress, "The Wood Eternal." Ground floor doors slide on Roll Rite Hangers, the highest priced hangers we handle. The haymow door in front gable is hung on three extra strong strap hinges. A smaller door on the mow floor level is convenient for many purposes. All doors are **"ready made"**—ready to hang in place and are superior to doors produced by hand carpentry.

Windows are four-light sash, opening size, 1 foot 10½ inches by 2 feet 9½ inches. In barns 24 feet wide the location of doors and windows is changed as shown in small illustration.

Hardware, such as latches, bolts, screws, nails, etc., is included in the prices, also sufficient paint for two coats, oxide red for the body and white for the trim, or you may select other combinations from our paint page. (See page 47.) Ventilating systems, cupolas, barn equipment, silos, foundation material, etc., are not included in the price.

Floor Plans. See pages 40 to 42 for sample floor plan arrangements. **Be sure to specify number of the floor plan that best suits your requirements.** Use the enclosed sketch sheet to outline your needs when changes are necessary and return for quotation on equipment.

Be sure to notice that all framing lumber is "already cut" and fitted, that doors are all **made and ready to hang,** that our roofing is guaranteed for fifteen years. **Take no substitute for Cypress Siding.**

QUALITY GUARANTEED

No. 2062

Door and Window Arrangement on Ground Floor of Barn 24 Feet Wide.

SIZE OF BARN			PRICE OF BARN		HAYMOW
Width Feet	Length Feet	Ht., From Sill to Eaves, Ft.	No. 1 Common Cypress Siding.	Select Cypress Siding	Capacity in Tons
24	28	12	$ 757.00	$ 796.00	18
24	34	12	853.00	896.00	22
24	40	12	948.00	996.00	26
24	46	12	1,045.00	1,095.00	29
24	52	12	1,140.00	1,195.00	33
28	34	12	1,009.00	1,057.00	30
28	40	12	1,123.00	1,176.00	35
28	46	12	1,237.00	1,293.00	40
28	52	12	1,351.00	1,410.00	46
28	58	12	1,454.00	1,528.00	51
30	34	12	1,094.00	1,152.00	33
30	40	12	1,216.00	1,278.00	39
30	46	12	1,339.00	1,405.00	45
30	52	12	1,462.00	1,532.00	51
30	58	12	1,585.00	1,659.00	57
30	64	12	1,708.00	1,798.00	62
30	70	12	1,831.00	1,912.00	68
30	76	12	1,954.00	2,052.00	74
32	34	14	1,153.00	1,216.00	40
32	40	14	1,282.00	1,349.00	47
32	46	14	1,412.00	1,934.00	54
32	52	14	1,540.00	1,616.00	61
32	58	14	1,669.00	1,762.00	68
32	64	14	1,798.00	1,883.00	75
32	70	14	1,927.00	2,023.00	82
32	76	14	2,056.00	2,151.00	89
32	82	14	2,184.00	2,284.00	96
32	88	14	2,314.00	2,405.00	103
34	34	16	1,288.00	1,370.00	49
34	40	16	1,425.00	1,513.00	57
34	46	16	1,560.00	1,654.00	66

"ALREADY CUT" AND FITTED. "ALREADY CUT" AND FITTED.

SIZE OF BARN			PRICE OF BARN		HAYMOW
Width Feet	Length Feet	Ht., From Sill to Eaves, Ft.	No. 1 Common Cypress Siding	Select Cypress Siding	Capacity in Tons
34	52	16	$1,698.00	$1,812.00	74
34	58	16	1,821.00	1,951.00	83
34	64	16	1,970.00	2,083.00	92
34	70	16	2,120.00	2,226.00	100
34	76	16	2,243.00	2,368.00	109
34	82	16	2,380.00	2,511.00	117
34	88	16	2,517.00	2,654.00	126
34	100	16	2,789.00	2,939.00	143
36	40	16	1,489.00	1,574.00	64
36	46	16	1,635.00	1,725.00	74
36	52	16	1,782.00	1,876.00	83
36	58	16	1,928.00	2,028.00	93
36	64	16	2,074.00	2,180.00	102
36	70	16	2,220.00	2,342.00	112
36	82	16	2,513.00	2,635.00	131
36	94	16	2,805.00	2,925.00	150
36	106	16	3,099.00	3,141.00	170
36	130	16	3,684.00	3,847.00	208
36	136	16	3,830.00	4,000.00	218
40	46	16	1,766.00	1,853.00	85
40	52	16	1,911.00	2,014.00	96
40	58	16	2,066.00	2,175.00	107
40	64	16	2,222.00	2,336.00	118
40	70	16	2,377.00	2,497.00	129
40	76	16	2,534.00	2,659.00	140
40	82	16	2,689.00	2,826.00	151
40	94	16	3,101.00	3,142.00	173
40	106	16	3,312.00	3,464.00	195
40	130	16	3,936.00	4,109.00	239
40	142	16	4,246.00	4,431.00	261

"ALREADY CUT" AND FITTED. "ALREADY CUT" AND FITTED.

What You Get

- All Wall Plates
- All Girders
- All Joists
- All Studding
- All Ridge Boards
- All Plates
- All Rafters
- All Trusses
- All Girts
- All Ribbons
- All Posts
- All Doors, Ready Made
- All Windows, Ready Made (Frames Ready Cut)

No. 1 Yellow Pine, "Already Cut" and Fitted.

See Prices on Opposite Page.

What You Get

- All Siding of Cypress, "The Wood Eternal."
- All Bridging.
- All Roof Boards.
- All Flooring.
- All Sub-Flooring.
- All Hardware.
- All Paint, enough for two coats.
- All Roofing, guaranteed for fifteen years.
- Free Building Plans.

High Quality.
Low Price.
Satisfaction
Guaranteed.

See Prices on Opposite Page.

Inexpensive Modern Bank Barns "Already Cut" and Fitted

Prices of the "Pocahontas" Modern Bank Barn No. 3005, "Already Cut" and Fitted.

SIZE OF BARN			PRICE OF BARN		HAY-MOW Capacity in Tons
Width, Feet	Length, Feet	Height From Sill to Eaves, Feet	With No. 1 Common Cypress Siding	With Select Cypress Siding	
30	32	16	$1,260.00	$1,316.00	45
30	42	16	1,492.00	1,567.00	59
30	46	16	1,586.00	1,653.00	55
30	56	16	1,819.00	1,895.00	69
30	60	16	1,912.00	1,990.00	73
30	70	16	2,144.00	2,231.00	87
30	74	16	2,237.00	2,328.00	103
30	84	16	2,471.00	2,568.00	117
30	88	16	2,564.00	2,665.00	123
30	98	16	2,796.00	2,905.00	136
32	32	16	1,306.00	1,365.00	51
32	42	16	1,546.00	1,614.00	67
32	46	16	1,642.00	1,714.00	73
32	56	16	1,882.00	1,962.00	89
32	60	16	1,978.00	2,061.00	96
32	70	16	2,218.00	2,310.00	112
32	74	16	2,314.00	2,409.00	118
32	84	16	2,555.00	2,657.00	134
32	88	16	2,650.00	2,757.00	141
32	98	16	2,890.00	3,005.00	157
34	42	16	1,607.00	1,678.00	75
34	46	16	1,707.00	1,783.00	82
34	56	16	1,960.00	2,044.00	100
34	60	16	2,061.00	2,147.00	108
34	70	16	2,313.00	2,409.00	126
34	74	16	2,414.00	2,513.00	134
34	84	16	2,667.00	2,773.00	152
34	88	16	2,768.00	2,878.00	158
34	98	16	3,020.00	3,139.00	176
36	42	16	1,673.00	1,748.00	80
36	46	16	1,776.00	1,855.00	88
36	56	16	2,036.00	2,123.00	107
36	60	16	2,139.00	2,230.00	115
36	70	16	2,399.00	2,497.00	134
36	74	16	2,503.00	2,604.00	141
36	84	16	2,762.00	2,871.00	160
36	88	16	2,865.00	2,979.00	167
36	98	16	3,125.00	3,247.00	186
36	112	16	3,488.00	3,621.00	213
40	42	16	1,767.00	1,842.00	92
40	46	16	1,875.00	1,960.00	101
40	56	16	2,146.00	2,240.00	123
40	60	16	2,256.00	2,352.00	132
40	70	16	2,527.00	2,632.00	154
40	74	16	2,635.00	2,743.00	163
40	84	16	2,908.00	3,033.00	185
40	88	16	3,016.00	3,135.00	194
40	98	16	3,287.00	3,415.00	215
40	112	16	3,668.00	3,807.00	246

This Description Applies to All Barns Priced on This and Illustrated on the Opposite Page.

THESE barns are designed after the most popular models in the United States.

The driveway—or barn floor—is 14 feet wide and provides ample room for threshing, baling of hay or storage. Additional driveways or barn floors can be arranged, if desired, at small cost.

These barns are to be erected on a foundation of concrete or masonry extending 8 feet 4 inches above the grade line. The ceiling height of the stable part, or ground floor, is 8 feet 6 inches.

All framing lumber is of No. 1 Yellow Pine.

The outside walls are sided with double V vertical siding of Cypress, "The Wood Eternal." This secures a very strong frame and outside walls that will withstand the weather a lifetime.

The floor in the stable part is to be of concrete or masonry. Flooring for the upper part consists of 1-inch flooring which is tongued and grooved to secure a better fit and to keep the ground floor free of dust from the mow above. In the driveway or barn floor, planks 2 inches thick are laid over the tongued and grooved flooring.

Prices of the "Big Chief" Modern Bank Barn No. 3022, "Already Cut" and Fitted.

SIZE OF BARN			PRICE OF BARN		HAY-MOW Capacity in Tons
Width, Feet	Length, Feet	Height From Sill to Eaves, Feet	With No. 1 Common Cypress Siding	With Select Cypress Siding	
30	32	16	$1,418.00	$1,484.00	57
30	42	16	1,676.00	1,749.00	75
30	46	16	1,778.00	1,843.00	82
30	56	16	2,036.00	2,121.00	100
30	60	16	2,139.00	2,227.00	108
30	70	16	2,396.00	2,493.00	126
30	74	16	2,499.00	2,598.00	134
30	84	16	2,757.00	2,864.00	152
30	88	16	2,923.00	2,971.00	158
30	98	16	3,117.00	3,235.00	176
32	32	16	1,482.00	1,552.00	64
32	42	16	1,752.00	1,829.00	84
32	46	16	1,859.00	1,939.00	92
32	56	16	2,127.00	2,215.00	112
32	60	16	2,234.00	2,327.00	120
32	70	16	2,503.00	2,602.00	140
32	74	16	2,610.00	2,713.00	148
32	84	16	2,880.00	2,990.00	168
32	88	16	2,986.00	3,101.00	176
32	98	16	3,254.00	3,377.00	196
34	42	16	1,808.00	1,889.00	92
34	46	16	1,919.00	2,003.00	101
34	56	16	2,182.00	2,287.00	123
34	60	16	2,305.00	2,400.00	132
34	70	16	2,571.00	2,684.00	154
34	74	16	2,692.00	2,798.00	162
34	84	16	2,968.00	3,082.00	185
34	88	16	3,078.00	3,196.00	193
34	98	16	3,354.00	3,480.00	215
36	42	16	1,882.00	1,967.00	101
36	46	16	1,999.00	2,087.00	110
36	56	16	2,294.00	2,389.00	135
36	60	16	2,411.00	2,510.00	145
36	70	16	2,705.00	2,812.00	168
36	74	16	2,823.00	2,933.00	178
36	84	16	3,116.00	3,235.00	203
36	88	16	3,234.00	3,356.00	212
36	98	16	3,528.00	3,659.00	236
40	42	16	2,007.00	2,100.00	117
40	46	16	2,127.00	2,222.00	128
40	56	16	2,425.00	2,529.00	152
40	60	16	2,542.00	2,651.00	163
40	70	16	2,841.00	2,956.00	187
40	74	16	2,961.00	3,080.00	198
40	84	16	3,258.00	3,386.00	222
40	88	16	3,378.00	3,509.00	233
40	98	16	3,676.00	3,815.00	257

Prices of the "Hiawatha" Modern Bank Barn No. 3023, "Already Cut" and Fitted.

SIZE OF BARN			PRICE OF BARN		HAY-MOW Capacity in Tons
Width, Feet	Length, Feet	Height From Sill to Eaves, Feet	With No. 1 Common Cypress Siding	With Select Cypress Siding	
30	32	16	$1,306.00	$1,366.00	48
30	42	16	1,539.00	1,609.00	63
30	46	16	1,634.00	1,705.00	69
30	56	16	1,867.00	1,947.00	79
30	60	16	1,959.00	2,045.00	90
30	70	16	2,194.00	2,286.00	105
30	74	16	2,288.00	2,383.00	111
30	84	16	2,521.00	2,625.00	126
30	88	16	2,614.00	2,735.00	132
30	98	16	2,849.00	2,964.00	147
32	32	16	1,356.00	1,421.00	57
32	42	16	1,605.00	1,675.00	75
32	46	16	1,700.00	1,778.00	82
32	56	16	1,946.00	2,032.00	100
32	60	16	2,044.00	2,133.00	108
32	70	16	2,290.00	2,387.00	126
32	74	16	2,389.00	2,489.00	134
32	84	16	2,635.00	2,744.00	158
32	88	16	2,733.00	2,844.00	176
32	98	16	2,979.00	3,099.00	61
34	42	16	1,660.00	1,737.00	80
34	46	16	1,762.00	1,843.00	88
34	56	16	2,019.00	2,108.00	106
34	60	16	2,121.00	2,213.00	114
34	70	16	2,376.00	2,477.00	133
34	74	16	2,480.00	2,583.00	141
34	84	16	2,735.00	2,858.00	160
34	88	16	2,838.00	2,953.00	168
34	98	16	3,094.00	3,218.00	186
36	42	16	1,723.00	1,843.00	84
36	46	16	1,830.00	1,912.00	92
36	56	16	2,093.00	2,223.00	112
36	60	16	2,198.00	2,292.00	120
36	70	16	2,461.00	2,602.00	140
36	74	16	2,567.00	2,673.00	148
36	84	16	2,829.00	2,982.00	168
36	88	16	2,935.00	3,052.00	176
36	98	16	3,198.00	3,362.00	196
40	42	16	1,861.00	1,950.00	101
40	46	16	1,973.00	2,065.00	110
40	56	16	2,252.00	2,353.00	134
40	60	16	2,364.00	2,468.00	144
40	70	16	2,643.00	2,755.00	168
40	74	16	2,755.00	2,870.00	178
40	84	16	3,034.00	3,158.00	201
40	88	16	3,146.00	3,273.00	211
40	98	16	3,425.00	3,560.00	235

To make the roof good and strong we furnish tight fitting tongued and grooved sheathing boards to be laid on the rafters and covered with Fire-Chief Shingle Roll Roofing, guaranteed by us for fifteen years. It resembles painted shingles.

Doors are made of Clear Cypress, "The Wood Eternal." For the basement of every barn quoted on this page we furnish four convenient Dutch Doors, No. 56A5060, 1⅝ inches thick. They can be located to suit the arrangement of the ground floor or stable. For the upper part at the entrance of the driveway, one pair of large Driveway Doors, No. 56A5059, opening size 12 feet by 12 feet, with a smaller built-in door. Opposite the driveway doors a smaller sliding door is provided, opening size 6 feet by 12 feet. Driveway doors slide on Roll Rite hangers. The small hinged door in one gable will be found very convenient. All doors are **"ready made"**—ready to hang in place.

Large windows 1⅜ inches thick with 9 lights, opening size 2 feet 8 inches by 3 feet 6 inches, are provided for one side, and the two ends of the ground floor. On the bank side two windows are furnished. All windows on the ground floor are furnished with ventilation shields. A quick change of air can be made by lifting the windows back into these shields. The practical construction of the shields protects your stock from direct drafts. For each gable we furnish three windows, 1⅜ inches thick with six lights, opening size 1 foot 9½ inches by 3 feet 6 inches.

Hardware, such as latches, bolts, screws, nails, etc., is included in the price, also sufficient paint for two coats—oxide red for the body and white for trim. You may select other combinations from our paint page. (See page 47.) Ventilating system, cupolas, foundation material and barn equipment are not included. Write for prices.

The floor plan for these barns can be arranged in the greatest variety, and without difficulty.

Our "already cut" feature, together with **"ready made"** doors, sash, etc., furnished at the price quoted, make it possible for you to build any of these barns at a remarkable saving of money, time and labor. Free building plans, which are easy to understand, are furnished with every order. Take no substitute for Cypress Siding. When ordering be sure to give the size.

ALREADY CUT AND FITTED PRICE $1,438.00 AND UP

A Modern Bank Barn, Trussed Roof Construction, Vertical Siding.

A HILLY countryside lends itself very favorably to this type of barn. It permits of easy access to the mow floor with team and load.

The basement of this barn is of concrete, brick, stone, etc., and is raised to the height of the mow floor. Ground floor, height to ceiling, 8 feet 6 inches. No foundation material for the basement part is included in the price. As most of these barns have an earthwall on at least one side of the basement, windows for ground floor are only furnished for the two ends and one side.

This barn is of the trussed roof type shown on page 5. The trusses are to be bolted together to secure greatest strength. Necessary bolts are included in the price.

All framing lumber is of No. 1 Yellow Pine.

The outside walls are covered with Double V vertical siding of Cypress, "The Wood Eternal." This combination secures a very strong frame and outside walls that will withstand the weather a lifetime.

To make the roof good and strong, we furnish **tight fitting tongued and grooved sheathing boards** to be laid on the rafters and covered with **Fire-Chief Shingle Roll Roofing. Fire-Chief Roofing** is guaranteed by us for **fifteen years.** It resembles painted shingles.

Doors are made of Clear Cypress, "The Wood Eternal." For the ground floor we furnish a pair of sliding doors for each gable. A pair of large driveway doors lead to the mow. One of the wings of the driveway doors is furnished with a small built-in door for personal entrance to the mow. (See page 43.) These doors slide on Roll Rite Hangers, the highest priced hangers we handle. The small hinged door in one gable will be found very convenient. All doors are **"ready made"**—ready to hang in place and are superior to doors produced by hand carpentry.

Large windows, 1⅜ inches thick, with nine lights; opening size, 2 feet 8 inches by 3 feet 6 inches for ground floor. Ventilation shields are furnished. A quick change of air can be made by lifting the windows back into the shields. The practical construction of the shields protects your stock from direct drafts. Three windows, six lights, 1⅜ inches thick, for each gable; opening size, 2 feet 8 inches by 3 feet 6 inches.

Hardware, such as latches, bolts, screws, nails, etc., is included in the price, also sufficient **paint for two coats,** oxide red for the body and white for the trim. You may select other combinations from our paint page (see page 47). Ventilating systems, cupolas, barn equipment, silos, foundation material, etc., are not included in the price.

Floor Plans. See pages 40 to 42 for sample floor plan arrangements. **Be sure to specify number of the floor plan that best suits your requirements.** Use the enclosed sketch sheet to outline your needs when changes are necessary and return for quotation on equipment.

Our "already cut" feature, together with **"ready made"** doors, sash, etc., furnished at the prices quoted, **make it** possible for you to get this modern bank barn at a remarkable saving of money, time and labor. **Free building plans,** which are easy to understand, are furnished with every order.

This barn will appeal to everyone who wants a good and solid bank barn.

Prices of the "Oneida" Modern Bank Barn No. 2063, "Already Cut" and Fitted.

SIZE OF BARN		Height From Sill to Eaves, Feet	PRICE OF BARN			SIZE OF BARN		Height From Sill to Eaves, Feet	PRICE OF BARN		
Width, Feet	Length, Feet		With No. 1 Common Cypress Siding ("Already Cut" and Fitted.)	With Select Cypress Siding ("Already Cut" and Fitted.)	Haymow Capacity in Tons	Width, Feet	Length, Feet		With No. 1 Common Cypress Siding ("Already Cut" and Fitted.)	With Select Cypress Siding ("Already Cut" and Fitted.)	Haymow Capacity in Tons
30	42	14	$1,438.00	$1,504.00	61	36	42	14	$1,683.00	$1,760.00	77
30	56	14	1,737.00	1,826.00	81	36	56	14	2,027.00	2,112.00	102
30	70	14	2,035.00	2,122.00	101	36	70	14	2,372.00	2,470.00	127
30	84	14	2,334.00	2,430.00	121	36	84	14	2,716.00	2,825.00	153
30	98	14	2,632.00	2,720.00	141	36	98	14	3,061.00	3,179.00	178
32	42	14	1,507.00	1,577.00	71	36	112	14	3,406.00	3,533.00	204
32	56	14	1,821.00	1,901.00	95	36	126	14	3,749.00	3,890.00	229
32	70	14	2,135.00	2,226.00	118	36	140	14	4,094.00	4,244.00	255
32	84	14	2,448.00	2,550;00	142	40	42	14	1,820.00	1,906.00	92
32	98	14	2,763.00	2,874.00	166	40	56	14	2,190.00	2,412.00	123
32	112	14	3,077.00	3,200.00	189	40	70	14	2,545.00	2,666.00	154
34	42	14	1,585.00	1,658.00	72	40	84	14	2,929.00	3,046.00	185
34	56	14	1,914.00	1,997.00	96	40	98	14	3,299.00	3,426.00	216
34	70	14	2,243.00	2,338.00	120	40	112	14	3,668.00	3,806.00	246
34	84	14	2,572.00	2,677.00	144	40	126	14	4,036.00	4,185.00	277
34	98	14	2,866.00	3,018.00	168	40	140	14	4,407.00	4,566.00	308
34	112	14	3,229.00	3,356.00	192	40	154	14	4,775.00	4,945.00	339
34	126	14	3,558.00	3,697.00	215	40	168	14	5,145.00	5,327.00	370

NOTE—Special door opposite driveway doors on mow floor, opening size 6 x 12 feet, including Roll Rite Barn Door Hangers and Track and barn door latch, complete, $17.00 extra.

ALREADY CUT AND FITTED PRICE $988.00 AND UP

A Modern Driveway Barn in Braced Rafter Construction, Drop Siding.

THIS Driveway Barn is much favored on account of its many conveniences. The arrangement permits driving into the barn with a load. The driveway may serve as a threshing floor or a place for implement storage in Winter. There is a big demand for this type of barn. Do not overlook the fact that the gambrel roof and the strong braced rafter construction permit of large storage capacity. Enough light is provided to make either section of the barn serve the purpose of a dairy barn or a barn for mixed stock. Where horses and cows are stabled under one roof, there is a possibility of a convenient floor arrangement to prevent the odors of the horse stable from mingling with the air in the cow stable. The feeding facilities also can be arranged very conveniently. Ground floor height, 8 feet 6 inches.

All framing lumber is of No. 1 Yellow Pine.

The outside walls are covered with drop siding of Cypress, "The Wood Eternal." This combination secures a very strong frame and outside walls that will withstand the weather a lifetime.

To make the roof good and strong, we furnish **tight fitting tongued and grooved sheathing boards,** to be laid on the rafters and covered with **Fire-Chief Shingle Roll Roofing. Fire-Chief Roofing** is guaranteed by us for **fifteen years.** It resembles painted shingles.

Doors are made of Clear Cypress, "The Wood Eternal." For the ground floor two pairs of spacious driveway doors, one pair on each side, and four sliding doors, two in each end are provided. One wing of the driveway doors is furnished with a smaller built-in door for personal entrance (see page 43). The doors slide on Roll Rite Hangers, the highest priced hangers we handle. Two smaller hinged doors, one in each end, at the mow floor level are convenient for many purposes. All doors are **"ready made"**—ready to hang in place and are superior to doors produced by hand carpentry.

Large nine-light windows, 1⅜ inches thick, opening size, 2 feet 8 inches by 3 feet 6 inches, are liberally provided for the ground floor. Ground floor windows are furnished with ventilation shields. A quick change of air can be made by lifting the windows back into the shields. The practical construction of the shields protects your stock from direct drafts.

Dormers over the driveway doors add materially to the good appearance of this barn.

In barns 24 feet wide, two 6-light windows are furnished for ends on ground floor, greater widths have 9-light windows.

Hardware, such as latches, bolts, screws, nails, etc., is included in the price, also sufficient **paint for two coats,** oxide red for the body and white for the trim. You may select other combinations from our paint page (see page 47). Ventilating systems, cupolas, barn equipment, silos, foundation material, etc., are not included in the price.

Floor Plans. See pages 40 to 42 for sample floor plan arrangements. **Be sure to specify number of the floor plan that best suits your requirements.** Use the enclosed sketch sheet to outline your needs when changes are necessary and return for quotation on equipment.

Our "already cut" feature, together with **"ready made"** doors, sash, etc., furnished at the prices quoted, make it possible for you to get this high class Driveway Barn at a remarkable saving of time and labor. Free building plans, which are easy to understand, are furnished with every order.

QUALITY GUARANTEED

Order now and save money. **Mention size when ordering.**

Prices of the "Fairview" Modern Driveway Barn No. 2068, "Already Cut" and Fitted.

SIZE OF BARN			PRICE OF BARN			SIZE OF BARN			PRICE OF BARN		
Width, Feet	Length, Feet	Height From Sill to Eaves, Feet	With No. 1 Common Cypress Siding	With Select Cypress Siding	Haymow Capacity in Tons	Width, Feet	Length, Feet	Height From Sill to Eaves, Feet	With No. 1 Common Cypress Siding	With Select Cypress Siding	Haymow Capacity in Tons
24	38	12	$ 988.00	$1.034.00	24	34	50	16	$1.632.00	$1.796.00	72
24	44	12	1.094.00	1.144.00	28	34	56	16	1.765.00	1.941.00	80
24	50	12	1.198.00	1.252.00	32	34	62	16	1.897.00	2.087.00	89
24	56	12	1.304.00	1.361.00	36	34	74	16	2.161.00	2.378.00	106
28	38	12	1.071.00	1.183.00	33	34	86	16	2.425.00	2.667.00	123
28	44	12	1.247.00	1.303.00	39	34	98	16	2.689.00	2.958.00	140
28	50	12	1.363.00	1.423.00	44	36	38	16	1.485.00	1.568.00	61
28	56	12	1.469.00	1.544.00	49	36	44	16	1.630.00	1.717.00	70
28	62	12	1.596.00	1.663.00	55	36	50	16	1.774.00	1.867.00	80
30	38	12	1.195.00	1.249.00	37	36	56	16	1.919.00	2.018.00	90
30	44	12	1.313.00	1.370.00	43	36	62	16	2.062.00	2.167.00	99
30	50	12	1.430.00	1.492.00	49	36	74	16	2.252.00	2.466.00	118
30	56	12	1.547.00	1.613.00	55	36	86	16	2.641.00	2.765.00	138
30	62	12	1.665.00	1.735.00	60	36	98	16	2.930.00	3.064.00	157
30	74	12	1.899.00	1.976.00	72	36	104	16	2.974.00	3.215.00	166
32	38	14	1.298.00	1.366.00	45	40	44	16	1.755.00	1.839.00	81
32	44	14	1.428.00	1.500.00	52	40	56	16	1.951.00	2.158.00	103
32	50	14	1.556.00	1.633.00	59	40	62	16	2,104.00	2.317.00	114
32	56	14	1.686.00	1.768.00	66	40	74	16	2.512.00	2.635.00	136
32	62	14	1.815.00	2.001.00	73	40	86	16	2.820.00	2.953.00	158
32	74	14	2.074.00	2.172.00	87	40	98	16	3.127.00	3.271.00	180
32	86	14	2.333.00	2.438.00	101	40	110	16	3.436.00	3.590.00	202
34	38	16	1.426.00	1.505.00	54	40	122	16	3.743.00	3.907.00	224
34	44	16	1.567.00	1.661.00	63	40	134	16	4.051.00	4.227.00	247

"Already Cut" and Fitted.

"Already Cut" and Fitted.

"Already Cut" and Fitted.

"Already Cut" and Fitted.

ALREADY CUT AND FITTED PRICE $690.00 AND UP

An Inexpensive Modern Barn With Large Driveway or Threshing Floor. "Already Cut" and Fitted.

HERE is a good, plain barn at a low price. It has a driveway, which can be used for threshing, baling of hay, storage of implements, etc. Both ends of the barn can be conveniently used as stables for horses or cattle. The driveway is open to the roof and the stables to both sides of the driveway are 8 feet 6 inches high. We believe that for a good, plain and serviceable barn this arrangement can hardly be improved upon. The strong, gambrel roof of this barn is self supporting and constructed as shown on page 4.

All framing lumber is of No. 1 Yellow Pine.

The outside walls are sided with drop siding of Cypress, "The Wood Eternal." This combination secures a very strong frame and outside walls that will withstand the weather for a lifetime.

To make the roof good and strong, we furnish tight fitting sheathing boards, to be laid on the rafters and covered with Fire-Chief Shingle Roll Roofing, guaranteed by us for fifteen years. It resembles painted shingles.

Doors are made of Clear Cypress, "The Wood Eternal." For the ground floor two pairs of spacious Driveway Doors No. 5059, one pair on each side of the barn are furnished, also four Dutch Doors No. 5060, 1⅜ inches thick, two in each end. For widths up to and including 30 feet only one of these Dutch doors for each end is furnished. The driveway doors are furnished with a smaller built-in door for personal entrance; they slide on Roll Rite hangers, the best hangers we handle. All doors are **"ready made"**—ready to hang in place and are superior to doors produced by hand carpentry.

Four light windows, 1⅜ inches thick, opening size 1 foot 10½ inches wide by 2 feet 9½ inches long, are provided for the ground floor.

Hardware, such as latches, bolts, screws, nails, etc., is included in the price, also sufficient paint for two coats, oxide red for the body and white for the trim. You may select other combinations from our paint page. (See page 47.) Ventilating systems, barn equipment and foundation material are not included in the price.

Floor Plans. See pages 40 to 42 for sample floor plan arrangements. **Be sure to specify number of the floor plan that best suits your requirements.** Use the enclosed sketch sheet to outline your needs when changes are necessary and return for quotation on equipment.

Our "already cut" feature, together with **"ready made"** doors, sash, etc., furnished at the prices quoted, make it possible for you to get this high class Colonial barn at a remarkable saving of time and labor. Free building plans, which are easy to understand, are furnished with every order.

Order now and save money.

Be sure to mention size when ordering.

Prices of the "Colonial" Modern Barn No. 3006, "Already Cut" and Fitted.

SIZE OF BARN			PRICE OF BARN		
Width, Feet	Length, Feet	Ht. From Sill to Eaves, Ft.	No. 1 Common Cypress Siding	Select Cypress Siding	Haymow, Capacity in Tons
24	24	12	$ 690.00	$ 723.00	16
24	30	12	782.00	818.00	21
24	36	12	877.00	914.00	25
24	42	12	966.00	1,010.00	29
24	48	12	1,058.00	1,106.00	34
24	54	12	1,150.00	1,201.00	39
28	30	12	865.00	903.00	27
28	36	12	970.00	1,012.00	32
28	42	12	1,075.00	1,122.00	37
28	48	12	1,179.00	1,230.00	42
28	54	12	1,284.00	1,349.00	48
30	30	12	907.00	948.00	30
30	36	12	1,021.00	1,065.00	36
30	42	12	1,133.00	1,183.00	42
30	48	12	1,247.00	1,300.00	48
30	54	12	1,361.00	1,417.00	54
30	60	12	1,474.00	1,535.00	60
32	36	14	1,142.00	1,203.00	42
32	42	14	1,271.00	1,337.00	50
32	48	14	1,387.00	1,469.00	58
32	54	14	1,529.00	1,592.00	65
32	60	14	$1,658.00	$1,736.00	72
32	66	14	1,786.00	1,869.00	79
32	72	14	1,915.00	2,003.00	86
34	36	16	1,223.00	1,291.00	50
34	42	16	1,357.00	1,435.00	58
34	48	16	1,492.00	1,575.00	67
34	54	16	1,628.00	1,715.00	75
34	60	16	1,763.00	1,855.00	84
34	66	16	1,898.00	1,996.00	92
34	72	16	2,034.00	2,136.00	101
34	78	16	2,168.00	2,276.00	110
36	36	16	1,303.00	1,368.00	58
36	42	16	1,433.00	1,513.00	68
36	48	16	1,573.00	1,658.00	77
36	54	16	1,713.00	1,803.00	87
36	60	16	1,852.00	1,947.00	96
36	66	16	1,992.00	2,093.00	105
36	72	16	2,133.00	2,238.00	114
36	78	16	2,272.00	2,383.00	124
36	84	16	2,399.00	2,528.00	134

ALREADY CUT AND FITTED PRICE $1,050.00 AND UP

A Modern Timber Frame Barn, "Already Cut" and Fitted.

IN ALL sections of the country we have found a great number of admirers of this type of barn. It embodies all the good features of framing which always have been the pride of the building craft. Our architects have used skill in designing a frame for this barn that combines the strength of the solid timber construction of the olden days with modern requirements.

The framework is built up of 2-inch material to the desired strength. Girder posts are solid. The simple way in which the framing is joined and the manner in which the bracing and trussing is done secure the admiration of every builder.

This barn can be built with considerably less lumber and far less labor than the old style timber frame barn. Its strength and rigidity will meet all requirements. Ground floor height, 9 feet.

All framing lumber is No. 1 Yellow Pine.

The outside walls are covered with double V vertical siding of Cypress, "The Wood Eternal." This combination secures a very strong frame and outside walls that will withstand the weather for a lifetime.

To make the roof good and strong we furnish **tight fitting tongued and grooved sheathing boards** to cover the rafters, on which you lay **Fire-Chief Shingle Roll Roofing. Fire-Chief Roofing** is guaranteed by us for **fifteen years.** It resembles painted shingles.

Doors are made of Clear Cypress, "The Wood Eternal." Two pairs of spacious driveway doors, one pair on each side, and four sliding doors, two in each end, are furnished for the ground floor. One wing of the driveway doors is furnished with a smaller built-in door for personal entrance. (See page 43.) The doors slide on Roll Rite Hangers, the highest priced hangers we handle. A smaller hinged door in each end at the mow floor level is convenient for many purposes. All doors are **"ready made"**—ready to hang in place and are superior to doors produced by hand carpentry.

Large nine-light windows, opening size, 2 feet 8 inches by 3 feet 6 inches, and 1⅜ inches thick, are liberally provided for the ground floor. They are furnished with ventilation shields. A quick change of air can be made by lifting the windows back into the shields. The practical construction of the shields protects your stock from direct drafts.

In barns 24 feet wide two six-light windows are furnished for the ends on ground floor.

Hardware, such as latches, bolts, screws nails, etc., is included in the price, also sufficient **paint for two coats,** oxide red for the body and white for the trim. You may select other combinations from our paint page (see page 47). Ventilating systems, cupolas, barn equipment, silos, foundation material, etc., are not included in the price.

Floor Plans. See pages 40 to 42 for sample floor plan arrangements. **Be sure to specify number of the floor plan that best suits your requirements.** Use the enclosed sketch sheet to outline your needs when changes are necessary and return for quotation on equipment.

Our "already cut" feature, together with **"ready made"** doors, sash, etc., furnished at the prices quoted, make it possible for you to get this modern timber frame barn at a remarkable bargain and a saving of time and labor. Free building plans, which are easy to understand, are furnished with every order.

QUALITY GUARANTEED

Prices of the "Prairie Chief" Modern Timber Frame Barn No. 2064, "Already Cut" and Fitted.

Size of Barn: Width, Feet	Length, Feet	Height From Sills to Eaves, Feet	Price of Barn: With No. 1 Common Cypress Siding	With Select Cypress Siding	Haymow, Capacity in Tons
24	42	14	$1,050.00	$1,146.00	29
24	56	14	1,283.00	1,340.00	39
28	42	14	1,217.00	1,269.00	34
28	56	14	1,486.00	1,547.00	45
28	70	14	1,768.00	1,822.00	56
30	42	16	1,322.00	1,384.00	42
30	56	16	1,622.00	1,694.00	56
30	70	16	1,921.00	2,005.00	70
30	84	16	2,221.00	2,327.00	84
30	98	16	2,520.00	2,627.00	98
32	42	16	1,399.00	1,462.00	50
32	56	16	1,704.00	1,778.00	67
32	70	16	2,008.00	2,093.00	84
32	84	16	2,313.00	2,407.00	101
32	98	16	2,618.00	2,724.00	118
32	112	16	2,921.00	3,038.00	134
34	42	18	1,459.00	1,533.00	63
34	56	18	1,786.00	1,874.00	84
34	70	18	2,114 00	2,202.00	105

Size of Barn: Width, Feet	Length, Feet	Height From Sills to Eaves, Feet	Price of Barn: With No. 1 Common Cypress Siding	With Select Cypress Siding	Haymow, Capacity in Tons
34	84	18	$2,441.00	$2,553.00	126
34	98	18	2,770.00	2,894.00	147
34	112	18	3,099.00	3,235.00	168
34	126	18	3,425.00	3,574.00	189
36	42	18	1,535.00	1,614.00	63
36	56	18	1,881.00	1,973.00	95
36	70	18	2,228.00	2,337.00	119
36	84	18	2,576.00	2,690.00	139
36	98	18	2,920.00	3,051.00	167
36	112	18	3,269.00	3,410.00	190
36	126	18	3,650.00	3,768.00	214
40	42	18	1,683.00	1,782.00	80
40	56	18	2,064.00	2,162.00	106
40	70	18	2,446.00	2,555.00	123
40	84	18	2,827.00	2,951.00	160
40	98	18	3,084.00	3,345.00	186
40	112	18	3,592.00	3,864.00	213
40	126	18	3,973.00	4,133.00	239
40	140	18	4,355.00	4,517.00	266

"Already Cut" and Fitted.

ALREADY CUT AND FITTED PRICE $1,216.00 AND UP

The "Economical Feeder" Modern Shed Barn. Greatest Floor Space at Lowest Cost. "Already Cut" and Fitted.

THIS modern barn with a lean-to, or shed on each side, is a desirable, low priced farm building, providing large floor space at a small expense.

The center of the barn is 8 feet 6 inches high to the lower edge of the joists, which are covered with tongued and grooved flooring, forming a large haymow above. No mow floor is provided for the sheds.

Solid posts and girders separate the center of the barn and the sheds. This gives a nearly unobstructed ground floor, which can be arranged to accommodate cattle, horses, sheep, etc.

All the framing lumber is of No. 1 Yellow Pine.

The outside walls are covered with drop siding of Cypress, "The Wood Eternal." This secures a very strong frame and outside walls that will last a lifetime.

To make the roof good and strong, we furnish tight fitting, tongued and grooved sheathing boards to be laid on the rafters and covered with Fire-Chief Shingle Roll Roofing, guaranteed by us for fifteen years. It resembles painted shingles.

Doors are made of clear Cypress, "The Wood Eternal." For the ground floor we furnish eight convenient Dutch doors; the haymow door in the front gable is hung on extra strong strap hinges. All doors are **"ready made"**—ready to hang in place and are superior to doors produced by hand carpentry.

Windows are four-light sash, opening size 1 foot 10½ inches by 2 feet 9½ inches.

Hardware, such as latches, bolts, screws, nails, etc., is included in the prices, also sufficient paint for two coats, oxide red for the body and white for the trim, or you may select other combinations from our paint page. (See page 47.) Ventilating systems, cupolas, barn equipment and foundation material are not included in the price.

The variety of sizes offered make possible a selection that will suit almost any requirements. In ordering be sure to specify the size of the barn and the sheds, as shown on this page.

The height of these barns includes the height of the foundation, which is to be 2 feet above the grade level, to keep moisture from the woodwork.

Prices of "Economical Feeder" Modern Shed Barn No. 3024, "Already Cut" and Fitted.

Width, Feet	Length, Feet	Width of Main Barn, Feet	Width of Each Shed, Feet	Height of Barn From Grade Line to Eaves, Feet	Greatest Height of Shed From Grade Line to Eaves, Feet	Price, With No. 1 Common Cypress Siding	Price, With Select Cypress Siding
44	42	20	12	14	12½	$1.216.00	$1.268.00
44	48	20	12	14	12½	1.336.00	1.392.00
44	54	20	12	14	12½	1.456.00	1.516.00
44	60	20	12	14	12½	1.576.00	1.640.00
44	66	20	12	14	12½	1.697.00	1.765.00
48	42	20	14	14	13	1.255.00	1.309.00
48	48	20	14	14	13	1.378.00	1.437.00
48	54	20	14	14	13	1.504.00	1.564.00
48	60	20	14	14	13	1.627.00	1.692.00
48	66	20	14	14	13	1.751.00	1.820.00
48	42	24	12	14	12½	1.327.00	1.384.00
48	48	24	12	14	12½	1.461.00	1.522.00
48	54	24	12	14	12½	1.594.00	1.660.00
48	60	24	12	14	12½	1.728.00	1.797.00
48	66	24	12	14	12½	1.861.00	1.933.00
52	48	24	14	14	13	1.505.00	1.566.00
52	54	24	14	14	13	1.642.00	1.708.00

Width, Feet	Length, Feet	Width of Main Barn, Feet	Width of Each Shed, Feet	Height of Barn From Grade Line to Eaves, Feet	Greatest Height of Shed From Grade Line to Eaves, Feet	Price, With No. 1 Common Cypress Siding	Price, With Select Cypress Siding
52	60	24	14	14	13	$1.774.00	$1.847.00
52	66	24	14	14	13	1.916.00	1.989.00
52	72	24	14	14	13	2.052.00	2.129.00
52	78	24	14	14	13	2.191.00	2.271.00
56	54	24	16	14	13½	1.699.00	1.766.00
56	60	24	16	14	13½	1.842.00	1.914.00
56	66	24	16	14	13½	1.985.00	2.058.00
56	72	24	16	14	13½	2.127.00	2.205.00
56	78	24	16	14	13½	2.265.00	2.350.00
56	84	24	16	14	13½	2.412.00	2.498.00
58	60	30	14	16	13	2.053.00	2.144.00
58	66	30	14	16	13	2.212.00	2.306.00
58	72	30	14	16	13	2.368.00	2.460.00
58	78	30	14	16	13	2,526.00	2.632.00
58	84	30	14	16	13	2.684.00	2.793.00
58	90	30	14	16	13	2.840.00	2.957.00
60	60	32	14	16	13	2.165.00	2.257.00

Width, Feet	Length, Feet	Width of Main Barn, Feet	Width of Each Shed, Feet	Height of Barn From Grade Line to Eaves, Feet	Greatest Height of Shed From Grade Line to Eaves, Feet	Price, With No. 1 Common Cypress Siding	Price, With Select Cypress Siding
60	66	32	14	16	13	$2.333.00	$2.427.00
60	72	32	14	16	13	2.502.00	2.605.00
60	78	32	14	16	13	2.671.00	2.779.00
60	84	32	14	16	13	2.841.00	2.954.00
60	90	32	14	16	13	3.010.00	3.128.00
62	60	30	16	16	13½	2.116.00	2.208.00
62	66	30	16	16	13½	2.275.00	2.377.00
62	72	30	16	16	13½	2.443.00	2.545.00
62	78	30	16	16	13½	2.606.00	2.712.00
62	84	30	16	16	13½	2.769.00	2.892.00
62	90	30	16	16	13½	2.932.00	3.050.00
64	60	32	16	16	13½	2.226.00	2.320.00
64	66	32	16	16	13½	2.402.00	2.501.00
64	72	32	16	16	13½	2.576.00	2.680.00
64	78	32	16	16	13½	2.752.00	2.861.00
64	84	32	16	16	13½	2.927.00	3.041.00
64	90	32	16	16	13½	3.101.00	3.220.00

Low Priced Modern "Pioneer" Barn No. 3008, "Already Cut" and Fitted.

Modern Shed or Lean-To No. 3003, "Already Cut" and Fitted.

SIZE OF BARN			PRICE OF BARN		Haymow Capacity, Tons
Width, Feet	Lgth., Feet	Height From Sill to Eaves, Feet	With No. 1 Common Cypress Siding	With Select Cypress Siding	
20	24	12	$ 598.00	$ 631.00	12
20	30	12	684.00	721.00	15
20	36	12	771.00	811.00	18
20	42	12	857.00	902.00	21
24	24	12	670.00	706.00	14
24	30	12	769.00	810.00	18
24	36	12	869.00	914.00	21
24	42	12	969.00	1,017.00	25
24	48	12	1,067.00	1,120.00	29
30	30	14	947.00	1,002.00	27
30	36	14	1,067.00	1,128.00	32
30	42	14	1,190.00	1,245.00	37
30	48	14	1,309.00	1,382.00	44
30	54	14	1,430.00	1,507.00	49
30	60	14	1,553.00	1,633.00	54
30	66	14	1,673.00	1,760.00	60
32	36	14	1,130.00	1,189.00	43
32	42	14	1,267.00	1,329.00	50
32	48	14	1,395.00	1,467.00	57
32	54	14	1,524.00	1,606.00	64
32	60	14	1,660.00	1,744.00	72
32	66	14	1,794.00	1,882.00	79
32	72	14	1,977.00	2,070.00	86
32	78	14	2,061.00	2,158.00	93

THE "Pioneer" Barn can be found in nearly all sections of the country. It is low in price and strongly constructed.

This barn is to be erected on a 2-foot concrete or masonry foundation. The ceiling is 8 feet 6 inches high.

All framing lumber is of No. 1 Yellow Pine.

The outside walls are sided with drop siding of Cypress, "The Wood Eternal." This combination secures a very strong frame and outside walls that will withstand the weather for a lifetime.

To make the roof good and strong, we furnish tight fitting tongued and grooved sheathing boards to be laid on the rafters and covered with Fire-Chief Shingle Roll Roofing, guaranteed by us for fifteen years. It resembles painted shingles.

Doors are made of Clear Cypress and are **"ready made."** Four Dutch doors and one gable hay door are furnished.

Four-light barn sash, opening size 1 foot 10½ inches by 2 feet 9½ inches, are furnished.

Hardware, such as latches, bolts, screws, nails, etc., is included in the price; also sufficient paint for two coats, oxide red for the body and white for the trim. You may select other combinations from our paint page. (See page 47.) Ventilating systems, cupolas, barn equipment, feed racks and foundation material not included.

Free building plans, which are easy to understand, are furnished with every order.

Be sure to give size when ordering.

AN ECONOMICAL way to increase the size of a barn is to add a shed or lean-to. We offer a variety of widths in such sheds at low prices.

They are designed to fit against the side of any barn, the wall of the barn to which they are to be attached to form the roof support.

Our sheds are to be erected on a 2-foot foundation of concrete or masonry to keep out the moisture. (Material for foundation not included in the price.) Foundation included in heights shown below.

All the framing lumber, roof boards, drop siding and window frames are **"already cut" and fitted.** Doors are **"ready made"**—ready to hang in place and are superior to those produced by hand carpentry.

Windows are four-light sash, opening size 1 foot 10½ inches by 2 feet 9½ inches. The roof is to be covered with Fire-Chief Shingle Roll Roofing, which resembles painted shingles, and is guaranteed by us for fifteen years.

All hardware and sufficient paint for two coats is also included, oxide red for the body and white for the trim. You may select other combinations from our paint page. (See page 47.)

We highly recommend these sheds as a valuable addition to your barn.

Width, Feet	Greatest Height From Grade Line, Feet	Height on Low Side From Grade Line, Ft.	Price, With No. 1 Common Cypress Siding		Price, With Select Cypress Siding	
			For Sheds 20 Feet Long	For Every Additional Foot in Length, ADD	For Sheds, 20 Feet Long	For Every Additional Foot in Length, ADD
12	12½	9	$146.00	$4.12	$152.00	$4.40
14	13	9	158.00	4.56	168.00	4.65
16	13½	9	171.00	4.94	181.00	5.10
18	14	9	183.00	5.32	195.00	5.50

ALREADY CUT AND FITTED PRICE $1,599.00 AND UP

A Modern Feeder's Barn, With a Gable Roof and Vertical Siding.

THIS Feeder Barn permits of a variety of uses. The large storage capacity makes it especially desirable where beef cattle are fattened during the Winter, to be sold in the Spring. The whole center of the barn can be filled with hay, roughage, etc., down to the ground. When this total space is not required, easy changes can be made for the storage of grain, etc., by building in bins or by flooring over part or all of the center of the barn. We make this barn in two widths, 60 feet and 52 feet. The interior provides for the feeding of loose cattle in the two side wings of the barn. Ground floor height in side wings, 8 feet 6 inches.

The center is separated from the side wings by a 5-foot partition. For the side wings a mow floor is furnished and a door at the level of the mow floor in each end. You will appreciate the convenience of this arrangement.

This barn is of the plank frame construction and has a straight gable roof.

All framing lumber is of No. 1 Yellow Pine.

The outside walls are covered with double V vertical siding of Cypress, "The Wood Eternal." This combination secures a very strong frame and outside walls that will withstand the weather for a lifetime.

To make the roof good and strong we furnish **tight fitting tongued and grooved sheathing boards** to be laid on the rafters and covered with **Fire-Chief Shingle Roll Roofing. Fire Chief Roofiing** is guaranteed by us for **fifteen years.** It resembles painted shingles.

Doors are made of Clear Cypress, "The Wood Eternal." Four pairs of sliding doors for the side wings are furnished, two pairs for each end of the barn, and one pair of large driveway doors, leading to the center of the barn. Driveway doors are furnished with built-in hinged door in one wing (see page 43). Over the driveway door a pair of sliding hay doors is furnished for one end. All these doors slide on Roll Rite Hangers, the highest priced hangers we handle. Four smaller hinged doors are furnished at the mow level over the sliding doors of the side wings. They are of great convenience. All doors are **"ready made"**—ready to hang in place—and are superior to doors produced by hand carpentry.

Four-light barn sash, opening size 1 foot 10½ inches by 2 feet 9½ inches, are furnished for the sides of the barn.

Hardware, such as latches, bolts, screws, nails, etc., is included in the price; also sufficient **paint for two coats,** oxide red for the body and white for the trim. You may select other combinations from our paint page (see page 47). Ventilating systems, cupolas, barn equipment, feed racks, silos, foundation material, etc., are not included in the price.

Floor Plans. See pages 40 to 42 for sample floor plan arrangements. **Be sure to specify number of the floor plan that best suits your requirements.** Use the enclosed sketch sheet to outline your needs when changes are necessary, and return for quotation on equipment.

Our "already cut" feature, together with **"ready made"** doors, sash, etc., furnished at the prices quoted, make it possible for you to get this high class Feeder Barn at a remarkable saving of time, money and labor. Free building plans, which are easy to understand, are furnished with every order.

This barn is a bargain at the prices quoted. **Be sure to give size when ordering.**

Prices of "Gable Feeder" Modern Feeder's Barn No. 2065, "Already Cut" and Fitted.

SIZE OF BARN			PRICE OF BARN		
Width, Feet	Length, Feet	Height From Sill to Eaves, Feet	With No. 1 Common Cypress Siding ("Already Cut" and Fitted.)	With Select Cypress Siding ("Already Cut" and Fitted.)	Hay Mow Capacity in Tons
52	40	12	$1,599.00	$1,654.00	60
52	48	12	1,788.00	1,858.00	72
52	56	12	1,957.00	2,053.00	84
52	64	12	2,166.00	2,247.00	96
52	72	12	2,357.00	2,542.00	108
52	80	12	2,547.00	2,637.00	120
60	56	12	2,234.00	2,324.00	115
60	64	12	2,441.00	2,537.00	131
60	72	12	2,648.00	2,751.00	148
60	80	12	2,856.00	2,962.00	164
60	88	12	3,063.00	3,174.00	180
60	96	12	3,270.00	3,389.00	197

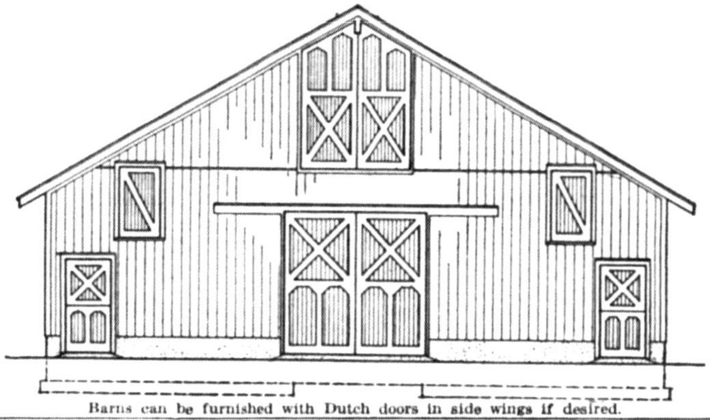

Barns can be furnished with Dutch doors in side wings if desired.

ALREADY CUT AND FITTED PRICE $1,269.00 AND UP

A Feeder's Barn, With a Gambrel Roof and Vertical Siding, "Already Cut" and Fitted.

THIS Feeder's Barn provides an abundance of loft room for the storage of hay, roughage, etc. The feature of the gambrel roof will appeal to you. The proportions of this Feeder's Barn are practical. They provide ample space for all requirements. This barn resembles the gable roof Feeder's Barn shown on the preceding page. Only high class material is furnished.

This barn is of the plank frame construction, and has a gambrel roof. The interior provides for the feeding of loose cattle in the two side wings of the barn. Ground floor height in side wings, 8 feet.

The center is separated from the side wings by a 5-foot partition. A mow floor is furnished for the side wings and there is a door at the level of the mow floor in each end. You will appreciate the convenience of this arrangement.

All framing lumber is of No. 1 Yellow Pine.

The outside walls are covered with double V vertical siding of Cypress, "The Wood Eternal." This combination secures a very strong frame and outside walls that will withstand the weather for a lifetime.

To make the roof good and strong, we furnish **tight fitting tongued and grooved sheathing boards,** to be laid on the rafters and covered with **Fire-Chief Shingle Roll Roofing. Fire-Chief Roofing** is guaranteed by us for **fifteen years.** It resembles painted shingles.

Doors are made of Clear Cypress, "The Wood Eternal." Four pairs of sliding doors for the side wings are furnished, two pairs for each end of the barn. In addition, one pair of large driveway doors, with built-in hinged door in one wing, (see page 43), leading to the center of the barn. Over the driveway door one pair of sliding hay doors is furnished. All these doors slide on Roll Rite Hangers, the highest priced hangers we handle. Four small hinged doors are furnished at the mow level over the sliding doors of the side wings. They are of great convenience. All doors are **"ready made"**—ready to hang in place and are superior to doors made by hand carpentry.

Four-light barn sash, opening size, 1 foot 10½ inches by 2 feet 9½ inches, are furnished for the sides of the barn.

Hardware, such as latches, bolts, screws, nails, etc., is included in the price, also sufficient **paint for two coats,** oxide red for the body and white for the trim. You may select other combinations from our paint page (see page 47). Ventilating systems, cupolas, barn equipment, feed racks, silos, foundation material, etc., are not included in the price.

Floor Plans. See pages 40 to 42 for sample floor plan arrangements. **Be sure to specify number of the floor plan that best suits your requirements.** Use the enclosed sketch sheet to outline your needs when changes are necessary and return for quotation on equipment.

Our "already cut" feature, together with **"ready made"** doors, sash, etc., furnished at the prices quoted, make it possible for you to get this high class Feeder's Barn at a remarkable saving of time and labor. **Free building plans,** which are easy to understand, are furnished with every order. They show exactly where every piece fits.

QUALITY GUARANTEED

Order Now and Save Money. Be Sure to Give Size!

Prices of "Gambrel Feeder" Modern Feeder's Barn No. 2066 "Already Cut" and Fitted.

SIZE OF BARN			PRICE OF BARN		
Width, Feet	Length, Feet,	Height From Sill to Eaves, Feet	With No. 1 Common Cypress Siding ("Already Cut" and Fitted)	With Select Cypress Siding ("Already Cut" and Fitted)	Hay-mow Capacity in Tons
52	40	10	$1,269.00	$1,322.00	67
52	48	10	1,419.00	1,475.00	80
52	56	10	1,569.00	1,627.00	93
52	64	10	1,718.00	1,779.00	106
52	72	10	1,867.00	1,932.00	119
52	80	10	2,017.00	2,085.00	132
60	56	10	1,783.00	1,847.00	126
60	64	10	1,951.00	2,019.00	144
60	72	10	2,119.00	2,191.00	162
60	80	10	2,288.00	2,363.00	180
60	88	10	2,456.00	2,534.00	198
60	96	10	2,624.00	2,706.00	216

Barns can be furnished with Dutch Doors in side wings if desired.

Prices of Modern Annex Barn No. 2067, "Already Cut" and Fitted.

SIZE OF BARN			PRICE OF BARN		SIZE OF BARN			PRICE OF BARN	
Width, Feet	Length, Feet	Height From Sill to Eaves, Feet	With No. 1 Common Cypress Siding	With Select Cypress Siding	Width, Feet	Length, Feet	Height From Sill to Eaves, Feet	With No. 1 Common Cypress Siding	With Select Cypress Siding
30	32	6	$398.00	$416.00	34	68	6	$ 724.00	$ 752.00
30	38	6	449.00	467.00	34	74	6	776.00	805.00
30	44	6	497.00	517.00	34	80	6	827.00	859.00
30	50	6	545.00	567.00	34	92	6	930.00	965.00
30	56	6	594.00	617.00	34	98	6	981.00	1,018.00
30	62	6	643.00	667.00	34	104	6	1,033.00	1,071.00
32	32	6	413.00	430.00	34	122	6	1,187.00	1,231.00
32	38	6	463.00	482.00	36	38	6	498.00	519.00
32	44	6	514.00	534.00	36	44	6	553.00	576.00
32	50	6	563.00	586.00	36	50	6	608.00	632.00
32	56	6	614.00	638.00	36	56	6	663.00	688.00
32	62	6	664.00	689.00	36	62	6	718.00	745.00
32	68	6	714.00	742.00	36	74	6	827.00	858.00
32	74	6	765.00	794.00	36	86	6	937.00	971.00
32	80	6	815.00	846.00	36	98	6	1,046.00	1,084.00
34	38	6	467.00	487.00	36	104	6	1,101.00	1,140.00
34	44	6	519.00	540.00	36	110	6	1,156.00	1,197.00
34	50	6	570.00	593.00	36	122	6	1,265.00	1,310.00
34	56	6	622.00	646.00	36	134	6	1,375.00	1,423.00
34	62	6	673.00	699.00					

"Already Cut" and Fitted.
"Already Cut" and Fitted.
"Already Cut" and Fitted.
"Already Cut" and Fitted.

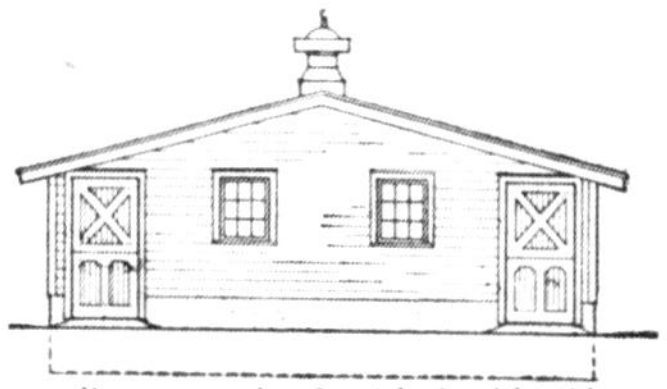

Barns can be furnished with side door arrangement if desired.

ALREADY CUT AND FITTED PRICE $398.00 AND UP

A Low Priced Modern One-Story Barn, "Already Cut" and Fitted.

THIS barn is designed to meet the requirements of an inexpensive **modern** dairy barn or a general purpose barn. As a sheep barn it is very practical. It may be used as an independent unit or built on to an old barn, as an "Annex." **This solves the problem of a modern sanitary addition at the least expense.** Only good quality of material is furnished for this barn.

All framing lumber is of No. 1 Yellow Pine.

The outside walls are covered with drop siding of Cypress, "The Wood Eternal." This combination secures a very strong frame and outside walls that will withstand the weather for a lifetime.

To make the roof good and strong, we furnish **tight fitting tongued and grooved sheathing boards** to be laid on the rafters and covered with **Fire-Chief Shingle Roll Roofing. Fire-Chief Roofing** is guaranteed by us for **fifteen years.** It resembles painted shingles.

Two pairs of sliding doors are provided, one pair for each end. They are made of Clear Cypress, "The Wood Eternal," and slide on Roll Rite Hangers, the most expensive hangers we handle. These doors are **"ready made"**—ready to hang in place and are superior to doors produced by hand carpentry.

Large nine-light windows, 1⅜ inches thick, opening size, 2 feet 8 inches by 3 feet 6 inches, provide plenty of light and air. Windows are equipped with ventilation shields. A quick change of air can be made by lifting the windows back into the shields. The practical construction of the shields protects your stock from direct drafts.

Hardware, such as latches, bolts, screws, nails, etc., is included in the price, also sufficient **paint for two coats,** oxide red for the body and white for the trim. You may select other combinations from our paint page (see page 47). Ventilating systems, cupolas, barn equipment, silos, foundation material, etc., are not included in the price.

Floor Plans. See pages 40 to 42 for sample floor plan arrangements. **Be sure to specify number of floor plan that best suits your requirements.** Use the enclosed sketch sheet to outline your needs when changes are necessary and return for quotation on equipment.

Be sure to notice that all our framing lumber, roof boards and siding are already cut and fitted, that doors are all **made and ready to hang,** that our roofing is guaranteed for fifteen years. **Take no substitute for Cypress Siding. When ordering be sure to give size.**

QUALITY GUARANTEED

Small Barns

WE ILLUSTRATE on this page three popular small barns. For low cost, solid construction and general appearance these barns are unexcelled.

The "Star" Modern Barn No. 3009, "Already Cut" and Fitted.

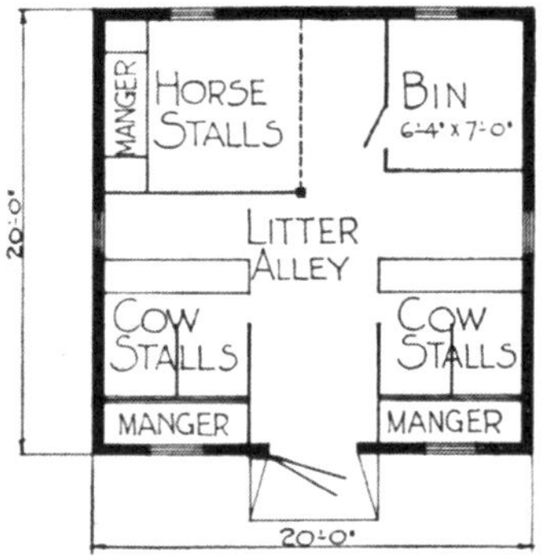

Floor Plan Suitable for Barn No. 3010.

The "Keystone" Modern Barn No. 3010, "Already Cut" and Fitted.

Prices of the "Star" Barn No. 3009, "Already Cut" and Fitted.

Width, Feet	Length, Feet	Height From Sills to Eaves, Feet	Price With No. 1 Common Cypress Siding	Price With Select Cypress Siding	Haymow Capacity, Tons
20	20	10	$304.00	$324.00	7
20	24	10	339.00	361.00	8½
20	28	10	373.00	397.00	10
20	32	10	407.00	433.00	11½

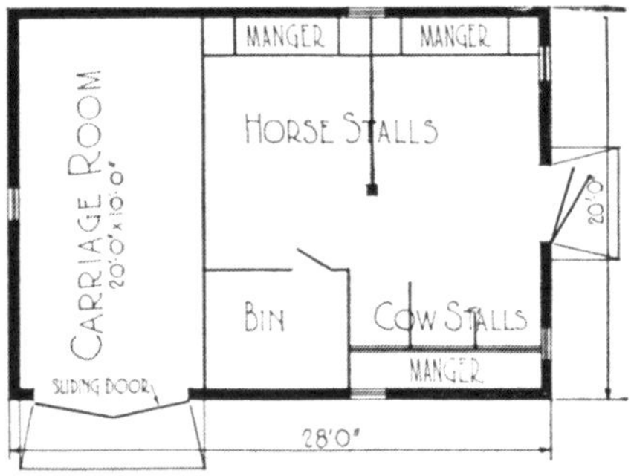

Floor Plan Suitable for Barns No. 3009 and No. 3011.

The "Crescent" Modern Barn No. 3011, "Already Cut" and Fitted.

Prices of the "Keystone" Barn No. 3010, "Already Cut" and Fitted.

Width, Feet	Length, Feet	Height From Sills to Eaves, Feet	Price With No. 1 Common Cypress Siding	Price With Select Cypress Siding	Haymow Capacity, Tons
20	20	10	$323.00	$346.00	8½
20	24	10	362.00	387.00	10
20	28	10	401.00	427.00	12½
20	32	10	440.00	467.00	14

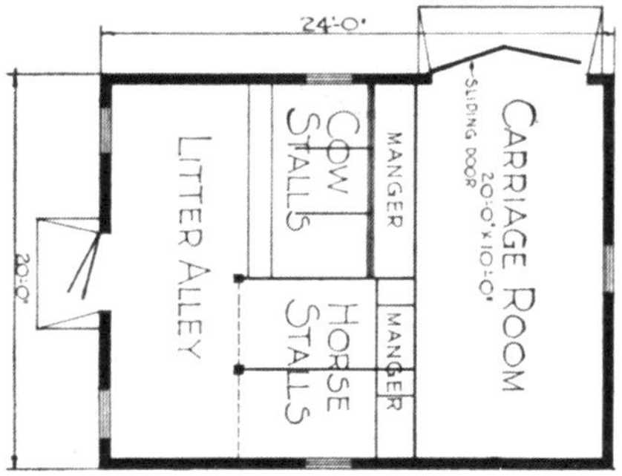

Floor Plan Suitable for Barns No. 3009 and No. 3011.

We offer each style in several lengths to meet most demands. Greater lengths quoted on request.

The floor arrangements suggested on this page can be changed to meet your requirements, and we will quote you a price including equipment on receipt of your sketch showing what you need.

All barns are to be erected on a foundation of concrete or masonry, extending 18 inches above the grade line. Ceiling height, 8 feet 6 inches.

Prices of the "Crescent" Barn No. 3011, "Already Cut" and Fitted.

Width, Feet	Length, Feet	Height From Sills to Eaves, Feet	Price With No. 1 Common Cypress Siding	Price With Select Cypress Siding	Haymow Capacity, Tons
20	20	10	$349.00	$369.00	10
20	24	10	391.00	411.00	12
20	28	10	432.00	453.00	14
20	32	10	473.00	495.00	16

WHAT YOU GET:

No. 1 Yellow Pine framing lumber; Cypress, "The Wood Eternal," drop siding for the outside walls; tight fitting sheathing boards for the roof, to be covered with Fire-Chief Shingle Roll Roofing, guaranteed by us for fifteen years; Clear Cypress doors, "ready made," and windows; all hardware and enough paint for two coats, which may be selected from our paint page. (See page 47.) Foundation material, interior equipment and cupolas are not included in the price.

All the lumber is **"already cut"** and fitted. Doors are **"ready made,"** ready to hang in place, all of which makes for a good strong barn and a saving in time, labor and money.

ALREADY CUT AND FITTED PRICE $1,627.00 AND UP

A Round Barn, 60 Feet in Diameter, "Already Cut" and Fitted

THE floor plans on this page indicate the variety of uses to which this barn may be put. It may be used either as an exclusive dairy barn or as a barn for general purposes. In fact, the possibilities of the arrangement of the floor plan are unlimited. You will appreciate the feature of the big driveway doors. Doors of most round barns are troublesome as they are usually cut and fitted on the job and bent to the radius of the barn. We have departed from the old style of a curved driveway door and furnish straight sliding doors that will give satisfaction. The illustration shows our method and the good appearance of the barn thereby obtained.

The barn is to be built on a 2-foot concrete foundation. Ground floor height, 8 feet 6 inches. Sills, plates, girders and purlins are furnished "already cut" and fitted, ready to be nailed up on the job. Height from sill to eaves, 16 feet.

All framing lumber is of No. 1 Yellow Pine.

The outside walls are covered with double V vertical siding of cypress, "The Wood Eternal." This combination secures a very strong frame and outside walls that will withstand the weather for a lifetime.

The roof is covered with **Extra Thick and Clear Red Cedar Shingles.**

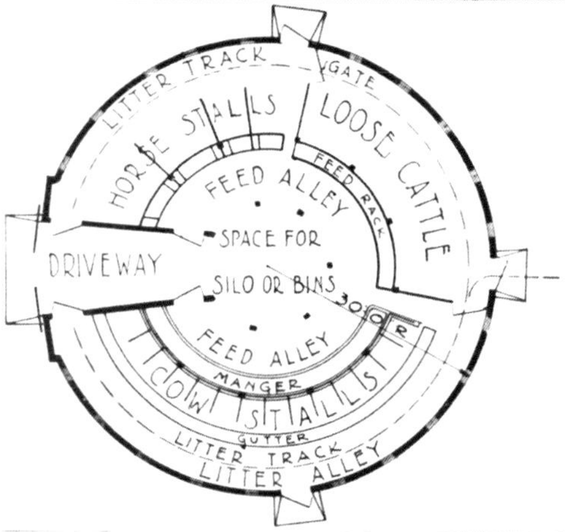

Prices of No. 2071 Round Barn, 60 Feet in Diameter, Total Height of Barn From Grade Line, 48 Feet. "Already Cut" and Fitted

Price, With No. 1 Common Cypress Siding $1,627.00
Price, With Select Cypress Siding 1,682.00

Doors are made of Clear Cypress, "The Wood Eternal." The large driveway doors slide on Roll Rite Hangers, the highest priced hangers we handle. In one of the wings a smaller hinged door for personal entrance is provided. Three more hinged doors for the entrance of stock are conveniently located. All doors are **"ready made"**—ready to hang in place and are superior to doors produced by hand carpentry.

Hinged windows, opening size, 1 foot 10½ inches by 2 feet 9½ inches, are furnished.

Hardware, such as latches, bolts, screws, nails, etc., is included in the price, also sufficient **paint for two coats,** oxide red for the body and white for the trim. You may select other combinations from our paint page (see page 47). Ventilating systems, cupolas, barn equipment, silos, foundation material, etc., are not included in the price.

Our "already cut" feature together with **"ready made"** doors at the prices quoted make this round barn a good investment, at a great saving of time and labor. Free building plans, which are easy to understand, are furnished with every order. They show exactly where every piece fits.

QUALITY GUARANTEED

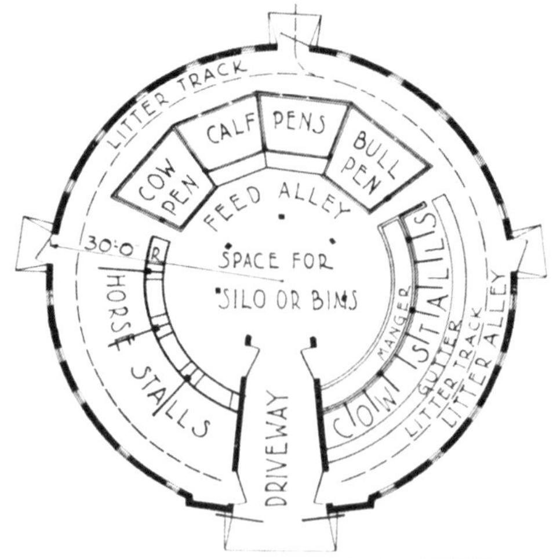

ALREADY CUT AND FITTED PRICE $2,123.00 AND UP

A Modern Octagon Barn of Popular Design, "Already Cut" and Fitted.

THIS Octagon Barn, like all other barns shown in this book, is of first class construction. It is becoming very popular throughout the country, because of the simplicity with which it may be erected and the great possibilities it offers for a practical floor arrangement. The floor plan on this page is only a suggestion of the convenience of this type of barn. It can be easily modified. Ground floor height, 8 feet 6 inches; height from sill to eaves, 18 feet.

All framing lumber is of No. 1 Yellow Pine.

The outside walls are covered with drop siding of cypress, "The Wood Eternal." This combination secures a very strong frame and outside walls that will withstand the weather for a lifetime.

To make the roof good and strong, we furnish **tight fitting, tongued and grooved sheathing boards,** to be laid on the rafters and covered with **Fire-Chief Shingle Roll Roofing. Fire-Chief Roofing** is guaranteed by us for **fifteen years.** It resembles painted shingles.

Doors are made of Clear Cypress, "The Wood Eternal." One large driveway door is provided for the entrance of team and load. This door slides on Roll Rite Hangers, the highest priced hangers we handle. One wing of the driveway doors is furnished with a small hinged door built in for personal entrance to the barn (see page 43). Three Dutch doors, conveniently located, are for the entrance of cattle. All doors are **"ready made"**—ready to hang in place and are superior to doors produced by hand carpentry.

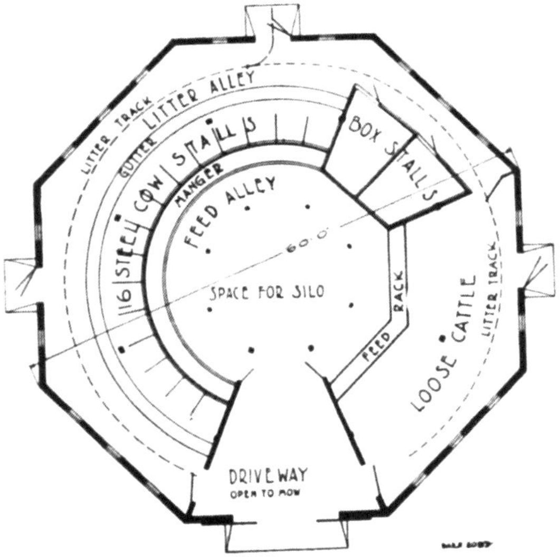

Prices of Our Modern "Octagon" Barn No. 2085, "Already Cut" and Fitted. Diameter, 60 Feet. Total Height of Barn From Grade Line, 44 Feet. Haymow Capacity, About 50 Tons.

Price, "Already Cut," With No. 1 Common Cypress Siding, "The Wood Eternal" $2,123.00
Price, "Already Cut," With Select Cypress Siding, "The Wood Eternal" 2,215.00

Large six-light windows, $1\frac{3}{8}$ inches thick, opening size, 1 foot $10\frac{1}{2}$ inches by 3 feet 6 inches, with ventilation shields, are furnished for the ground floor. A quick change of air may be made by lifting the windows back into the shields. The practical construction of the shields protects your stock from direct drafts.

Light for the mow is provided by a four-light sash in every alternate panel of the cupola and one six-light sash, which can be opened when filling the silo. Between the lights in the cupola slatted shutters for mow ventilation are furnished.

Hardware, such as latches, bolts, screws, nails, etc., is included in the price, also sufficient **paint for two coats,** oxide red for the body and white for the trim. You may select other combinations from our paint page (see page 47). Ventilating systems, barn equipment, silos, concrete, etc., are not included in the price.

Our "already cut" feature, together with **"ready made"** doors, sash, etc., furnished at the prices quoted, make it possible for you to get this Modern Octagon Barn at a remarkable bargain, as well as a saving of time and labor. Free building plans, which are easy to understand, are furnished with every order. **Be sure to give size when ordering your barn.**

QUALITY GUARANTEED

Modern L Shaped Barn, "Already Cut" and Fitted. Composed of Nos. 2054 and 2060. See Pages 7 and 13.

L SHAPED BARNS are favored for many reasons. Feeding and all chores may be done from one central place. Different kinds of stock may be housed close together, and individual surroundings may be maintained. The illustration shows the joining of two of our barns, in form of an L. The main barn, our No. 2054, arranged as a dairy barn; the wing, our No. 2060, arranged for horses. Both cows and horses are close together, but the interior arrangement is such that the air of the cow barn does not mingle with that of the horse barn, and feeding facilities can be easily arranged to secure greatest convenience.

Nearly all our barns can be arranged in a similar manner. The cost of an L shap barn is exactly the cost of the two sizes you select from this catalog.

We have priced a few exceptionally good sizes below. If you do not find one them suitable to your requirements, you will have no difficulty in selecting the desir combination from this catalog.

Be sure to notice that all our framing lumber is **"already cut" and fitted,** that doors are **"ready made"** ready to hang in place, and that our roofing is guaranteed for fifteen years. **Take no substitute for Cypr Siding.**

Prices of L Shaped Barns.

FOR MAIN BARN					FOR SIDE WING					
No.	Width of Barn, Feet	Length of Barn, Feet	Height From Sill to Eaves, Feet	Price, with No. 1 Common Cypress Siding	No.	Width of Barn, Feet	Length of Barn, Feet	Height From Sill to Eaves, Feet	Price, with No. 1 Common Cypress Siding	Price of Complete L Shaped Barn, With No. 1 Common Cypress Siding
2054	36	110	16	**$3,577.00**	**2060**	32	56	14	**$1,676.00**	**$5,253.00**
2054	34	86	16	**2,794.00**	**2060**	30	50	12	**1,495.00**	**4,289.00**
2054	30	50	12	**1,643.00**	**2060**	24	32	12	**856.00**	**2,498.00**

ALREADY CUT AND FITTED PRICE $345.00 AND UP

The Famous "Sunlight" Hog House, "Already Cut" and Fitted.

THIS is a good hog house designed to provide suitable quarters for Winter farrowing. The young pigs can be turned into the early pastures, and will soon be ready for an early market. The difference in the prices obtainable is well known to the average breeder, which consideration has prompted us to adopt and recommend our "Sunlight" Hog House. It fully deserves its name. It permits the sunshine around March 1st, the usual time of Winter farrowing, to reach the pens on both sides of the building.

This hog house is to be built with its ridge running East and West. On the South side two rows of large roof windows permit of a nearly continuous flow of sunshine on the floor of the pens. In the South wall, a third row of windows is provided, directly opposite the nests.

The outside pen doors slide up and down, and may be easily opened by pulling a rope from the center aisle. This convenience saves many a step and the unpleasant task of walking into the pens or outside of the hog house in bad weather.

If pen doors are wanted for rear side, add $1.50 each.

We furnish this type of hog house 24 feet wide and 20 feet wide. Twenty-four feet wide hog houses have double sliding doors at each end, which are hung on Roll Rite Hangers, the best we handle, and lead to an alley which is 8 feet wide. Pens to be 8 feet deep and 6 feet wide. The hog house 20 feet wide provides for one single sliding door at each end hung on Roll Rite Hangers. Here the alley is only 4 feet wide.

All framing lumber is of No. 1 Yellow Pine.

The outside walls are covered with drop siding of Cypress, "The Wood Eternal." This combination secures a very strong frame and outside walls that will withstand the weather for a lifetime.

To make the roof good and strong, we furnish **tight fitting tongued and grooved sheathing boards** to be laid on the rafters and covered with **Fire-Chief Shingle Roll Roofing. Fire-Chief Roofing** is guaranteed by us for **fifteen years.** It resembles painted shingles.

Hardware, such as latches, bolts, screws, nails, etc., is included in the price, also sufficient **paint** for outside walls **for two coats,** oxide red for the body and white for the trim. You may select other combinations from our paint page (see page 47). Pen partitions are not included in the price. On page 46 you will find desirable pen partitions which are especially recommended.

Be sure to notice that all our framing lumber is already cut and fitted, that doors are all **made and ready to hang,** that our roofing is guaranteed for fifteen years. **Take no substitute for cypress siding. Be sure to give size when ordering Hog House.**

Prices of Modern "Sunlight" Hog Houses No. 264A2072 and No. 264A2073 "Already Cut" and Fitted.

No.	Size of Hog House		Prices, "Already Cut" and Fitted	
	Width	Length	With No. 1 Common Cypress Siding ("Already Cut" and Fitted)	With Select Cypress Siding ("Already Cut" and Fitted)
264A2073	20	24	$345.00	$359.00
264A2073	20	30	409.00	423.00
264A2073	20	36	472.00	489.00
264A2073	20	42	536.00	554.00
264A2073	20	48	600.00	619.00
264A2073	20	54	662.00	685.00
264A2073	20	60	727.00	751.00
264A2072	24	24	$437.00	$454.00
264A2072	24	30	506.00	525.00
264A2072	24	36	576.00	596.00
264A2072	24	42	645.00	668.00
264A2072	24	48	714.00	739.00
264A2072	24	54	785.00	809.00
264A2072	24	60	854.00	880.00
264A2072	24	72	992.00	1,021.00

ALREADY CUT AND FITTED PRICE $556.00 AND UP

An Inexpensive and Practical Hog House, "Already Cut" and Fitted.

THE "Community" Hog House keeps your hogs warm and comfortable in the Wintertime because the space allowed for each animal is limited and on account of the storage room overhead. It is to be located with the ridge running north and south, which, in connection with tight fitting doors and the absence of windows on the north end of the ground floor, excludes the cold north winds. In Summertime this hog house is remarkably cool.

The building is to be erected on a 10-inch foundation of concrete or masonry. The width is 22 feet; height to the ceiling, 6 feet 6 inches, and the total height, 18 feet. Over the ceiling is sufficient room for the storage of bedding, hay and feed. **"Already cut"** material for stairs, 2 feet wide, leading to the storage room is furnished. In the northwest corner of the building a space 6½ feet wide is reserved for a feeding and boiler room. A bin for the storage of grain is located directly over the feed room. This bin is 4 feet wide, 6 feet long and 4 feet high. All framing lumber is of No. 1 Yellow Pine. The outside walls are covered with No. 1 Common **Cypress drop siding,** "The Wood Eternal."

To make the roof good and strong, we furnish **tight fitting sheathing boards** to be laid on the rafters and covered with **Fire-Chief Shingle Roll Roofing,** guaranteed by us for fifteen years. This roofing resembles painted shingles.

Doors. A Dutch door, No. 56A5059, size, 3 feet 9 inches by 6 feet 6 inches, is furnished for one end; also two smaller doors, 3 feet by 4 feet, for the mow floor at each end, giving access from the outside to the storage room (see illustration). Doors for the pens provide for an opening 22 inches wide and 32 inches high. They slide up and down and by means of a rope may be opened from the feed alley. This saves you the trouble of entering the pen or going outside of the hog house on a rainy day. All doors are **"ready made"**—ready to hang in place and are superior to doors made by hand carpentry.

Prices of Modern Community" Hog House. No. 3004, "Already Cut" and Fitted.

No.	Width, Feet	Length, Feet	Prices, "Already Cut" and Fitted	No.	Width, Feet	Length, Feet	Prices, "Already Cut" and Fitted
3004	22	26	$556.00	3004	22	44	$846.00
3004	22	32	659.00	3004	22	50	945.00
3004	22	38	753.00				

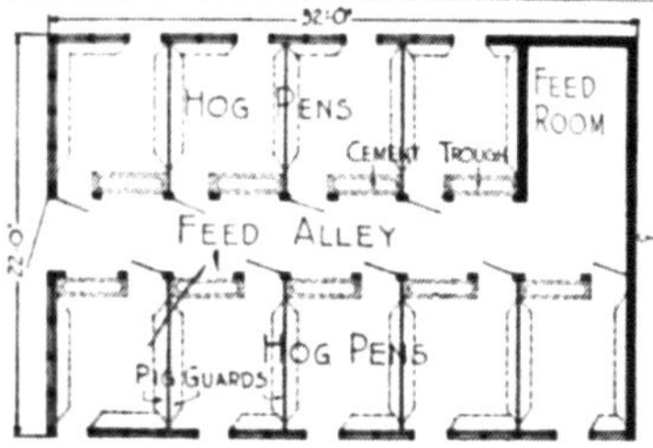

The above prices include the following interior equipment: (1) All pen partitions. (2) All front panels and hardware, as described on page 46. (3) Overhead flooring. (4) Stairs. (5) Grain bin. (6) Solid feed room partition. (7) Pig fenders. Do not overlook this when considering prices.

Floor plan showing arrangement of "Community" Hog House 32 feet long.

Windows. We furnish our General Barn Sash, No. 63A5082, which is 1⅜ inches thick, four lights, 9 inches by 14 inches each; opening size, 1 foot 10½ inches by 2 feet 9½ inches. These windows are placed close together and allow an abundance of sunshine to enter the building (see illustration). The north end has only one window—in the gable.

Hardware, such as latches, bolts, screws, nails, etc., is included in the price, also sufficient **paint for two coats,** oxide red for the body and white for the trim. You may select other combinations from our paint page (see page 47). Cupolas, ventilating systems and concrete or masonry are not included.

Pens are 6 feet wide and 9 feet deep, outside measurements. They extend to both sides of the feed alley, which is 4 feet wide. Material for the necessary pen partitions and pig fenders is furnished.

This "Community" Hog House is recommended for cold climates by the Nebraska Agricultural State College, many other authorities and prominent breeders. It was the only permanent model among the exhibits at the Chicago Stock Show in December, 1917.

QUALITY GUARANTEED

Be sure to notice how complete is the equipment of our Hog Houses as specified above, that all framing lumber is "already cut" and fitted, that doors are all "ready made" and ready to hang, that all roofing is guaranteed for fifteen years, and all bins and pen partitions included in price. Be sure to give the size when ordering the hog house.

ALREADY CUT AND FITTED PRICE $306.00 AND UP

The "Monitor" Hog House, "Already Cut" and Fitted.

MOST people are familiar with this style of hog house and know that it is a dependable type. We introduce only such styles of hog houses as have proved serviceable. The "Monitor" Hog House has some very good features.

It should be built with the ridge running **east and west to get a southern exposure,** which will permit the sun's rays to penetrate every nook and corner. This applies during the early Springtime, when sunshine is equally as necessary as feed for the Winter farrowing. In Summer, when the sun rises high in the sky, the hottest rays of the noontide will little affect the interior of a hog house of this type. Instead, during the hot spell of the year, the hog house will be a welcome retreat for your swine during midday hours, for the perpendicular rays of the sun reflect but little upon the windows of the structure.

The floor arrangement provides for pens 8 feet deep and 6 feet wide, outside measure.

All framing lumber is of No. 1 Yellow Pine.

The outside walls are covered with drop siding of Cypress, "The Wood Eternal," which secures a very strong frame and outside walls that will withstand the weather for a lifetime.

To make the roof good and strong we furnish **tight fitting tongued and grooved sheathing boards** to be laid on the rafters and covered with **Fire-Chief Shingle Roll Roofing. Fire-Chief Roofing** is guaranteed by us for **fifteen years.** It resembles painted shingles.

Prices of Modern "Monitor" Hog Houses, No. 2074 and No. 2075, "Already Cut" and Fitted.

No.	Size of Hog House: Width, Feet	Length, Feet	Total Height From Grade Line, Feet	Prices, "Already Cut" and Fitted: With No. 1 Common Cypress Siding	With Select Cypress Siding
2074	20	24	14	**$306.00**	**$322.00**
2074	20	30	14	**365.00**	**382.00**
2074	20	36	14	**424.00**	**436.00**
2074	20	42	14	**472.00**	**503.00**
2074	20	48	14	**542.00**	**563.00**
2074	20	54	14	**601.00**	**625.00**
2074	20	60	14	**659.00**	**684.00**
2074	20	72	14	**776.00**	**785.00**
2075	24	24	14	**354.00**	**371.00**
2075	24	30	14	**417.00**	**435.00**
2075	24	36	14	**479.00**	**500.00**
2075	24	42	14	**542.00**	**563.00**
2075	24	48	14	**604.00**	**627.00**
2075	24	54	14	**666.00**	**690.00**
2075	24	60	14	**728.00**	**754.00**
2075	24	72	14	**854.00**	**881.00**

"Already Cut" and Fitted. **"Already Cut" and Fitted.**

Doors are made of Cypress, "The Wood Eternal." They are **"ready made"**—ready to hang in place and are superior to doors produced by hand carpentry. A pair of hinged doors at each end are furnished for size 24 feet wide; a single hinged door at each end for size 20 feet wide. Pen doors are made to slide up and down and may be raised from the inside by a rope. Hog House No. 2074 has 4-foot alley and one swinging door. No. 2075 has 8-foot alley and double swinging doors.

Hardware, such as latches, bolts, screws, nails, etc., is included in the price, also sufficient paint for outside walls for **two coats,** oxide red for the body and white for the trim. You may select other combinations from our paint page (see page 47). Pen partitions are not included in the price. On page 46 you will find desirable pen partitions which are especially recommended.

Our **"already cut"** feature, together with **"ready made"** doors, sash, etc., furnished at the prices quoted, make it possible for you to get this Modern "Monitor" Hog House at a remarkable bargain as well as a saving of time and labor. If pen doors are wanted for rear side, add $1.50 each.

QUALITY GUARANTEED

IMPORTANT

Our No. 1 Common Cypress Siding will outwear Hemlock, Spruce, Fir or White Pine. Select Cypress Siding is more durable than any other wood for siding.

"Wigwam" Movable Hog House No. 56A2076
Size, 8 Feet Long, 6 Feet Wide.
Price, "Already Cut" and Fitted, with Fire-Chief Shingle Roll Roofing and Paint, each $41.60
The same, without Fire-Chief Shingle Roll Roofing and without Paint, each $34.30
Hardware Included.

"Colony" Hog House No. 56A2077
Size, 8 Feet Long, 6 Feet Wide.
Price, "Already Cut" and Fitted, with Fire-Chief Shingle Roll Roofing and Paint, each $38.48
The same, without Fire-Chief Shingle Roll Roofing and without Paint, each $33.76
Hardware Included.

The advantages of a movable hog house are many. They are especially desirable for the isolation of swine with their litter. They also permit of pasturing the herd at distant places from the general farmstead, providing necessary comfort at such places and eliminating much work otherwise connected with the chores.

Two types of movable hog houses are especially recommended, the "Wigwam" Hog House and the "Colony" Hog House.

The "Wigwam" Hog House

THIS type is of triangular shape. It is built up on two solid 4x4 timbers or runners. The runners serve as a sleigh on which the whole hog house is moved from place to place. The total length of the hog house is 8 feet; its width, 6 feet; height, 7 feet. All the material employed for the "Wigwam" is No. 1 yellow pine. Each side of the "Wigwam" has two doors, extending over the whole width of structure. These doors may be opened by means of a wooden prop hinged right to the door. This permits of adjustment of doors to meet weather requirements. The floor is made of 2-inch planks. A fixed pig fender runs across both ends of the "Wigwam." The outside is covered with shiplap siding. On the gables the siding is not brought right to the top. A small space is left open in order to secure ventilation when "Wigwam" is closed.

The "Colony" Hog House

OUR "Colony" Hog House differs from the "Wigwam" in that it provides more overhead room. This is because it is built like a house with a gable roof. The outside dimensions of this type are 6 feet wide by 8 feet long. Its greatest height is a little over 5 feet. It rests like the "Wigwam" on two solid runners, 4x4 inches. All the lumber is No. 1 yellow pine. The floor of the hog house is made of 2-inch planks. The "Colony" is provided with three doors, two forming the roof and one forming the side. All doors are hung on strap hinges. Another small door for ventilation is fixed in one end. For the end walls fixed fenders are furnished. For the side walls, removable fenders. The outside is covered with shiplap siding. On the gable the "Colony" is not sheathed right to the ridge, a small space being left open for ventilation.

A Modern Corn Crib and Granary, "Already Cut" and Fitted

THIS is a combination corn crib and granary. The difference between this building and the corn cribs shown on page 39 of this catalog is that Nos. 2081 and 2082 are long and low, while No. 2084, shown at the right, is high and short. The result is, that on account of the saving in cost of foundation and roof material, the comparative cost per bushel capacity is less. Besides, the specifications of this building provide for a stronger structure, necessitated on account of the grain bins.

This combination corn crib and granary is very highly recommended. The main feature is the strength of the structure, designed to carry with safety **any kind** of grain that may be stored in the bins. Occasionally a builder specifies the capacity of the bins for a certain number of bushels of **"oats"** or other light grain. The bins in this granary are strong enough to carry either oats, wheat or flax seed with safety. Our bins are estimated for a safe carrying load of over 75 pounds per bushel capacity. The safe carrying load required for oats is only 32 pounds per bushel capacity.

Our corn crib and granary is 28 feet wide. The driveway is 12 feet wide and 12 feet high to the lower edge of the joists. This permits of the safe dumping of wagon loads with bat boards. The corn cribs are each 8 feet wide and have a capacity of over 51 bushels per running foot. The grain bins are 12 feet wide and have a capacity of 77 bushels of wheat per running foot.

The floor of each bin is hopper shaped and fitted with two grain spouts with shut-offs.

Modern Corn Crib and Granary No. 2084.

ALREADY CUT AND FITTED PRICE $1,066.00 AND UP

The foundation of this building should be of concrete, 18 inches above the grade line. A shelling and vent trench may be easily built under the floor of the cribs. The total **height** of the building from the grade to the eaves **is 17 feet 6 inches,** and to the peak 32 feet. **The** roof is crowned with a wooden cupola. **This** permits of the installation of an inside cup **elevator.** The construction is such, that an **outside** elevator may be employed with equally good results.

All framing lumber, cribbing and partitions used in this building are No. 1 Yellow Pine; all drop siding Cypress, "The Wood Eternal."

To make the roof good and strong, we **furnish tight fitting, tongued and grooved sheathing boards** to be laid on the rafters and covered with **Fire-Chief Shingle Roll Roofing. Fire-Chief Roofing** is guaranteed by us for **fifteen years.** It resembles painted shingles.

The big driveway doors are "ready made" of Cypress, "The Wood Eternal." They slide on Roll Rite Hangers, the best hangers we handle. One pair of doors is provided for each gable. A smaller door is built in one of the wings of these driveway doors. This permits easy personal access to the interior. (See page 43.) At each end of the cribs hinged doors of cribbing are specified. These doors are not ready made, because they can be better fitted on the job.

Specially designed openings are arranged along the inside driveway to be used for filling and emptying the cribs. These openings are adjustable.

Light for the bins is provided by two hinged sash in the cupola and one stationary sash in each gable.

Hardware, such as latches, bolts, screws, nails, etc., is included in the price, also sufficient **paint for two coats,** for outside of the building, oxide red for the body and white for the trim. You may select other combinations from our paint page (see page 47). Foundation material, equipment, elevators, etc., are not included in the price.

Our "already cut" feature, together with **"ready made"** sliding doors, sash, etc., furnished at the prices quoted, make it possible for you to get this Modern Corn Crib and Granary at a remarkable saving of money, time and labor. Building plans, which are easy to understand, are furnished with every order.

We highly recommend this Modern Corn Crib and Granary.

Prices of No. 2084, "Already Cut" and Fitted.

SIZE			CAPACITY		Price
Width Feet	Length, Feet	Height From Sill to Eaves, Feet	Of Corn Cribs, About Bushels	Of Grain Bins, About Bushels	
28	28	16	3,190	2,156	**$1,066.00**
28	34	16	3,865	2,618	**1,207.00**
28	40	16	4,550	3,080	**1,350.00**
28	46	16	5,235	3,542	**1,491.00**
28	52	16	5,916	4,004	**1,633.00**

The capacity of cribs and bins has been estimated very conservatively.

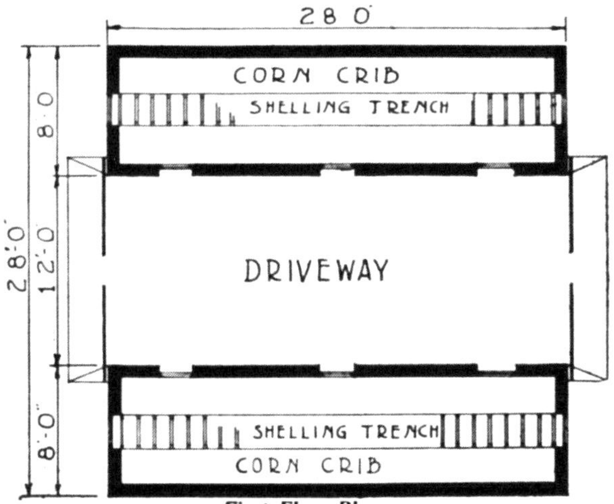

First Floor Plan.

Second Floor Plan.

Modern Multiple-Unit Poultry House No. 2078, "Already Cut" and Fitted

ALREADY CUT AND FITTED PRICE $226.00
Price Does Not Include Concrete.

20x20 feet. Price, each $242.00

Additional Section, 20x20 feet. Price, each $226.00

THIS Modern Multiple-Unit Poultry House is the result of the most careful study, made by the New Jersey Agricultural Experiment Station. It is designed to solve the question of the poultry man who is striving for a good Winter egg yield, and it has been approved by the poultry departments of a large number of agricultural colleges. The Multiple-Unit Poultry House comes as near to the ideal as is possible under average conditions.

A house 20 feet wide and 20 feet long, of the shed roof type, is one of the most economical units that can be used.

Such a unit has a floor space of 400 square feet, allowing 4 square feet to each bird. It is adapted to all climates and is one of the most efficient producing units. The diagrams below show the possible development of the Multiple-Unit idea, by the addition of more units, as may be desired, to increase the size of the plant.

This Modern Multiple-Unit Poultry House is to be built on an 8-inch concrete foundation. It is 9 feet high in front, 5 feet high in the rear. An entrance is provided at one end. Two sash, opening size 2 feet 8½ inches by 4 feet 7½ inches, are furnished for each unit, also two muslin screens, 3 feet by 4 feet 6 inches. The sash is provided with hinges to swing sideways. The screens have hinges to be swung upward if desired. The roosts **ARE INCLUDED** in price. They are hinged to the side wall studs and can be swung up, to allow the cleaning of the drop board. Material to build 21 nest boxes is furnished. Each box to be 12 inches wide, 16 inches long and 18 inches deep. Material for one coop for brooding hens also furnished. All framing lumber is No. 1 yellow pine.

For the outside walls drop siding is furnished. For the roof we furnish tight fitting tongued and grooved sheathing boards, on which is to be laid **Fire-Chief Shingle Roll Roofing,** guaranteed by us for **fifteen years.**

This Multiple-Unit Poultry House should be built facing the South on a layer of cinders and with concrete foundation walls, to keep rats and vermin out. If you follow these instructions you will have a well ventilated Laying House, free from all forms of moisture, well provided with direct sunlight, built to keep the birds warm, and large enough to permit plenty of exercise. All framing for this house is ready cut, all you have to do is to nail them together according to our plans, which are easy to understand and furnished with every order.

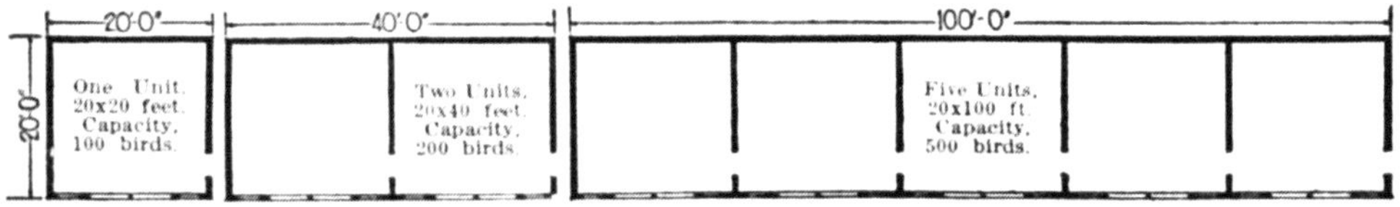

Modern Chicken House No. 2051, "Already Cut" and Fitted

FLOOR PLAN

ALREADY CUT AND FITTED PRICE $331.00
Concrete Not Included.

For $331.00 we will furnish all the material to build this Chicken House, consisting of Lumber, Siding, Windows, Doors, Hardware, Paint and Wire Netting.

On looking at the floor plans it will be seen that entrance can be made from either end of the chicken house directly into the feed alley and from the alley into any one of the pens. These are four in number, each one measuring 13 feet 6 inches by 7 feet 6 inches. Wire netting, 6 feet high around each pen. The dimensions of the building are 28 feet in length by 20 feet in width.

Poultry raising has been demonstrated to be quite a profitable business and the most noted experts in this line agree that a well lighted, well ventilated chicken house is a necessity where the biggest profits are desired. Especially is this true of people who make a specialty of select prize stock. In this Chicken House we provide twelve big windows, six on each side. In addition to these windows we further provide four four-light windows which swing outward at the top of the building, all of same being on the front. No chance for draft if windows are left open, yet this arrangement gives an abundance of light and fresh air. For the roof we specify our 4-Ply Best-of-all Roofing, GUARANTEED FOR TWELVE YEARS.

Build this Modern Chicken House on a concrete foundation 6 to 10 inches above the ground, 16 to 24 inches deep, 4 to 6 inches wide. Fill inside of foundation with dry cinders and put 2-inch mixture of concrete on top of this. This will give you a dry and sanitary chicken house, easy to clean, relatively cheap and permanent.

Modern Sanitary Dairy House No. 2086, "Already Cut" and Fitted.

ALREADY CUT AND FITTED PRICE $628.00 CYPRESS.

THE appearance of this modern and sanitary Milk and Dairy House will appeal to you particularly if you look at the practical and labor saving arrangement shown in the floor plan.

The floor is of concrete, sloping to the center to permit of an easy off-flow. The foundation walls are carried 3 feet above the floor line or 5 feet above the grade line. Two outside platforms are provided, one for loading the finished product, located near the cooler or ice box, the other for receiving the empty bottles, on the side of the washroom. These platforms are about 2 feet above the grade line. The main entrance is reached by a concrete walk, which permits of easier access to the dairy house than steps. The height of the building from the finished floor to the ceiling is 9½ feet, the full height is 19 feet. The building is 30 feet long and 22 feet wide.

All framing lumber is of No. 1 yellow pine.

The outside walls are covered with drop siding of Cypress, "The Wood Eternal." This combination secures a very strong frame and outside walls that will withstand the weather for a lifetime.

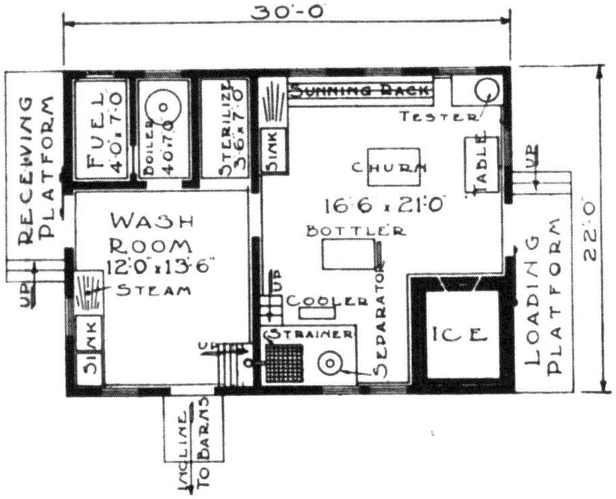

To make the roof good and strong, we furnish **tight fitting, tongued and grooved sheathing boards,** which are to be laid on the rafters and covered with **Fire-Chief Shingle Roll Roofing. Fire-Chief Roofing** is guaranteed by us for **fifteen years.** It resembles painted shingles.

Doors are "ready made" at our factory and are better doors than could be made by hand carpentry. The main entrance door is a strong, five-panel hinged door. Doors at the receiving and loading platforms slide on Roll Rite hangers, the highest priced hangers we handle.

Spacious windows provide light and air.

Ceiling and inside walls are covered with beaded ceiling, ½x4-inch No. 1 yellow pine.

Hardware, such as latches, bolts, screws, nails, etc., is included in the price, also sufficient **paint for two coats,** dark gray house paint for the body and white for the trim. You may select other combinations from our paint page (see page 47). Foundation material, interior equipment, cupolas, and paint for the interior are not included in the price. This dairy house is a paying investment at a moderate price.

Milk House No. 2093, "Already Cut" and Fitted.

ALREADY CUT AND FITTED PRICE $240.00

For $240.00 we will furnish all the material to build this Milk House, consisting of Lumber, Siding, Flooring, Shingles, Hardware, Door, Sash and Paint.

THE Milk House above illustrated is a solidly constructed building, 14 feet wide by 12 feet long and is 16 feet high from foundation to peak. It has two water troughs in which the milk cans can be set to cool until they are ready to be taken to the creamery or until the milk is disposed of otherwise. On the left hand side a shelf runs from the door to the side wall and extends along the side wall to the water trough in the rear.

To the farmer who wants to keep his milk in a sanitary condition, this building at the price which we charge for all the materials is a first class proposition. We guarantee a good grade of material all the way through and agree to furnish a sufficient quantity to build it according to our plans. No. 1 yellow pine framing lumber. Cement, dairy equipment, etc., not included in the price.

A Good Modern Ice House, Strongly Built and Well Insulated. "Already Cut" and Fitted

THIS ice house is carefully and solidly constructed. It is not what you put into an ice house that counts, but what you take out of it. Ice in many ice houses will shrink to about 50 per cent of the original filling. If the ice is well packed, shrinkage will be reduced to a minimum; we estimate that it should be less than 20 per cent, which we believe is about the least shrinkage on record.

The walls of this ice house are substantially insulated (see cross sections of the ground floor at the bottom of this page). The construction and insulation are guaranteed to give satisfactory results.

All siding and partitions used for this ice house are of Cypress, "The Wood Eternal," a wood that will not rot. We could have reduced the cost considerably by substituting other lumber which is far less expensive, but that would not have given us the permanent building which we are offering you. The ice house can be furnished with a built-in cold storage room, as is shown in the illustration in the lower left hand corner.

ALREADY CUT AND FITTED PRICE $637.00 AND UP

The U. S. Department of Agriculture says:*

"An ample supply of ice is of greater economic importance in the country home than in the city residence. City people can purchase perishable supplies as needed, but the remoteness of country homes from markets often renders it necessary to use canned, corned or smoked meat products during the season of the year when the table should be supplied with fresh meats. Not only is ice appreciated because of its use in the preservation of fresh meats, butter and other table supplies, but the production of high grade domestic dairy products is almost impossible without it. Many markets to which milk is now shipped demand that it be cooled before shipment to a degree not attainable without the use of ice.

"The principles involved in the proper care of milk and cream are very simple and easily understood. Notwithstanding this fact, much of the product delivered to creameries is improperly cared for and therefore is unfit to be made into butter. One of the most common causes of poor quality of butter is the lack of immediate, thorough cooling of the cream after separation. The Dairy Division has made a careful investigation of conditions existing on a large number of dairy farms where first class cream is produced, and the data obtained show that, if properly cooled, cream of the best grade can be produced with but little extra labor or expense.

"These dairymen realize the importance of the use of ice and provide themselves with a suitable supply."

*Farmers' Bulletins Nos. 475 and 623.

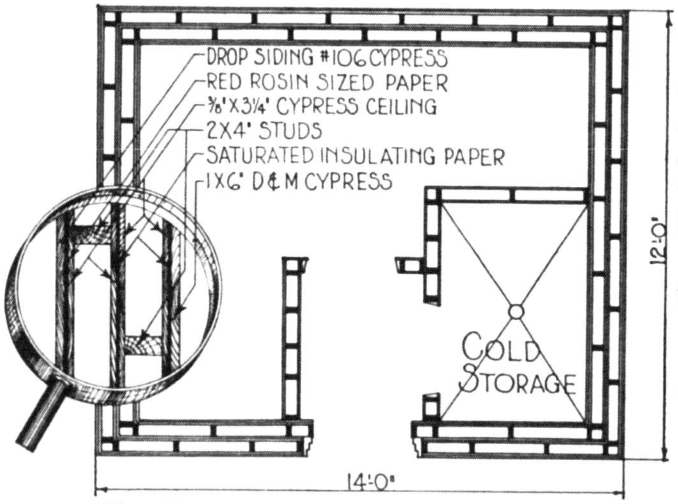

Floor Plan of Ice House With Cold Storage Room.

The "Alaska" Modern Ice House No. 3012, "Already Cut" and Fitted.

All the framing lumber for this ice house comes to you **"already cut" and fitted,** which not only reduces the building expense but, due to accurate fitting, makes tight joints, very essential in preventing the melting of the ice. Free building plans which are easy to understand are furnished with every order. An investment in an ice house of this kind will soon pay for itself.

Don't build an ice house with single studding and improperly insulated.

Prices of The "Alaska" Modern Ice House No. 3012, "Already Cut" and Fitted.

Width Feet	Length, Feet	Capacity, Tons	Price Without Cooling Room	Price With Cooling Room
14	12	42	**$637.00**	**$659.00**
14	16	58	**734.00**	**758.00**
14	20	75	**831.00**	**858.00**

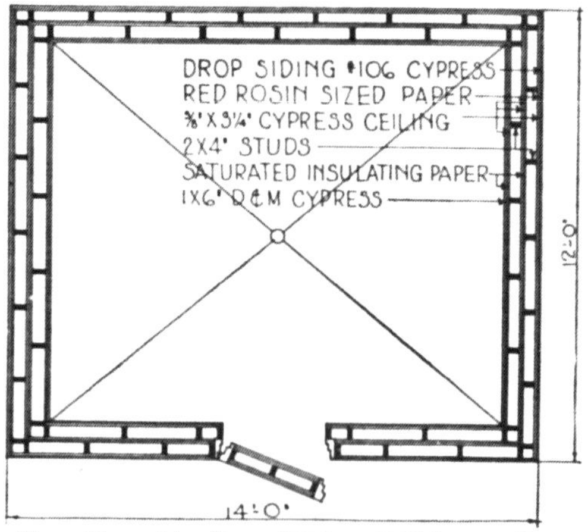

Floor Plan of Ice House Without Cold Storage Room.

Modern Corn Crib No. 2081

THE corn cribs offered on this page are of correct design and solid construction. We furnish them in two widths. The illustration shows our No. 2081, which is 26 feet wide. This building provides for a center driveway of 10 feet and two cribs, on each side of the driveway, each crib 8 feet wide. The same arrangement applies to Corn Crib No. 2082, but here the width of the cribs is only 6 feet each.

The cribbing is beveled on edges and spaced far enough apart to permit good circulation of air. This construction of the double crib embodies all the good features of a single crib which is wholly surrounded by air. But it is far cheaper than two single cribs, and, in addition, more convenient. On the inside of the cribs, doors of cribbing are provided for filling. These doors are not ready made, as they can be easily made on the job. Openings are also provided for the emptying of the crib.

The illustration shows the solid construction of this crib. The material furnished is first class in every respect.

All framing lumber is of No. 1 Yellow Pine.

Outside walls are covered with Yellow Pine Cribbing.

To make the roof good and strong, we furnish **tight fitting tongued and grooved sheathing boards,** to be laid on the rafters. On the sheathing boards is laid **Fire-Chief Shingle Roll Roofing. Fire-Chief Roofing** is guaranteed by us for **fifteen years.** It resembles painted shingles.

Hardware is included in the price, also sufficient **barn paint for two coats,** oxide red being specified. Foundation material is not included in prices.

Prices of Modern Corn Cribs No. 2081 and No. 2082, "Already Cut" and Fitted.

Catalog No.	Width, Feet	Length, Feet	Height From Sill to Eaves, Feet	Capacity (bushels)	Price
2081	26	30	10	2,066	**$445.00**
2081	26	36	10	2,560	**514.00**
2081	26	42	10	2,988	**579.00**
2081	26	48	10	3,412	**646.00**
2081	26	54	10	3,840	**713.00**
2082	22	24	10	1,280	**318.00**
2082	22	30	10	1,602	**381.00**
2082	22	36	10	1,920	**444.00**
2082	22	42	10	2,238	**508.00**
2082	22	48	10	2,560	**575.00**

"Already Cut" and Fitted.

Implement Shed No. 2088

FARMERS everywhere recognize the advisability of having a separate building for machinery and tools. Much money is wasted by not taking proper care of the machines which are required on the average farm throughout the year. The rain rusts them quickly and otherwise impairs their usefulness. When housed in a building of this kind they are always fully protected. The building above will also make a handy workshop for necessary repairs on all kinds of machinery, vehicles, harness and other odd jobs.

All the framing lumber is of No. 1 Yellow Pine.

The outside walls are covered with drop siding of Cypress, "The Wood Eternal." This combination secures a very strong frame and outside walls that will withstand the weather for a lifetime.

To make the roof good and strong we furnish **tight fitting tongued and grooved sheathing boards** to be laid on the rafters. On the sheathing boards is laid **Fire-Chief Shingle Roll Roofing. Fire-Chief Roofing** is guaranteed by us for **fifteen years.** It resembles painted shingles.

Doors are made of Clear Cypress, "The Wood Eternal." They are 10 feet high and 12 feet wide, built in two sections, and slide on Standard Barn Door Hangers and roller tracks. All doors are **"ready made"**—ready to hang in place, and are superior to doors produced by hand carpentry.

Windows are conveniently located.

Hardware, such as latches, bolts, screws, nails, etc., is included in the price, also sufficient **paint for two coats,** oxide red for the body and white for the trim. You may select other combinations from our paint page. (See page 47.) Cupolas, foundation material, etc., are not included in the price.

Prices of Implement Shed No. 2088, "Already Cut" and Fitted.

SIZES			PRICES	
Width, Feet	Length, Feet	Height to Peak, Feet	With No. 1 Common Cypress Siding	With Select Cypress Siding
24	24	17	$ 422.00	$ 447.00
24	36	17	554.00	580.00
24	48	17	682.00	712.00
24	60	17	812.00	845.00
24	72	17	942.00	978.00
24	84	17	1,073.00	1,110.00

"Already Cut" and Fitted. **"Already Cut" and Fitted.**

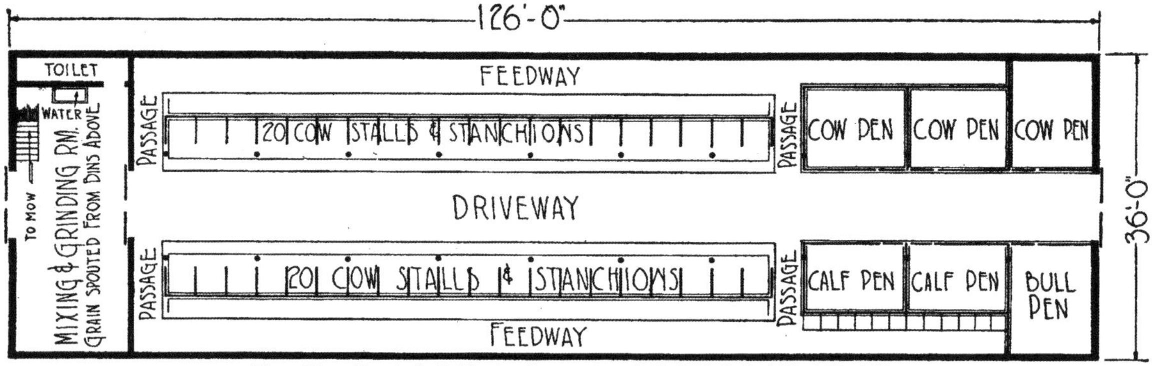

Floor Plan No. 1, Suitable for Barns Nos. 2055, 2057, 2058, 2059, 2063, 2067.

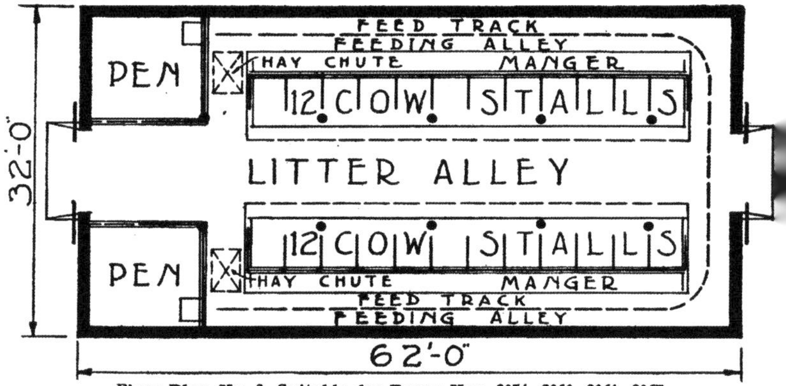

Floor Plan No. 3, Suitable for Barns Nos. 2054, 2060, 2061, 2067.

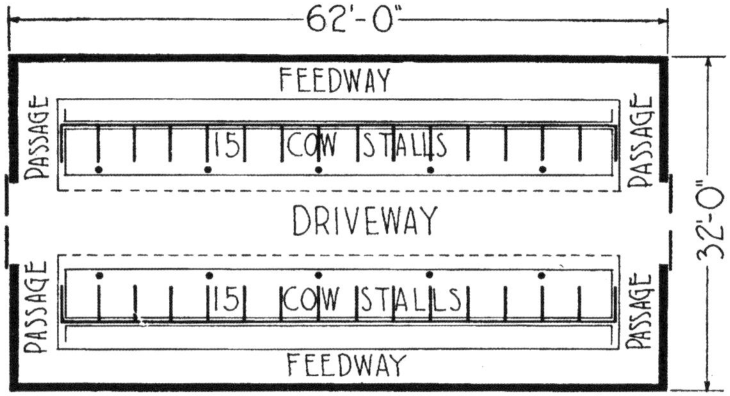

Floor Plan No. 4, Suitable for Barns Nos. 2054, 2060, 2061, 2067.

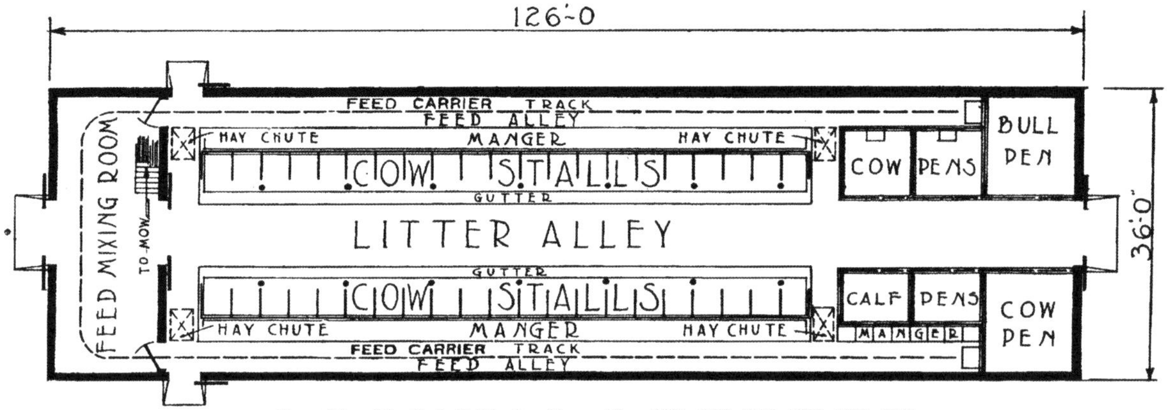

Floor Plan No. 2, Suitable for Barns Nos. 2055, 2057, 2058, 2059, 2063, 2067.

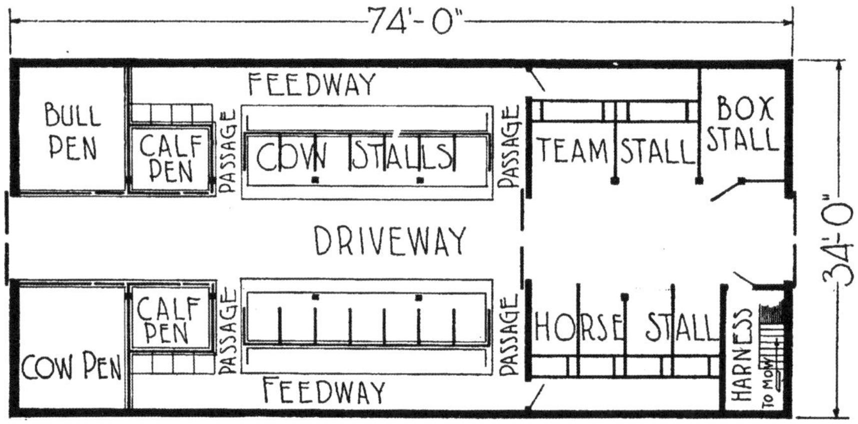

Floor Plan No. 5, Suitable for Barns Nos. 2054, 2060, 2061, 2067.

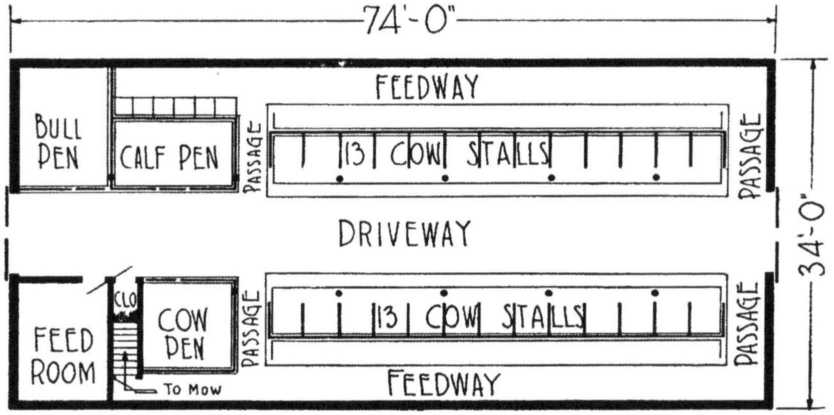

Floor Plan No. 6, Suitable for Barns Nos. 2054, 2060, 2061, 2067.

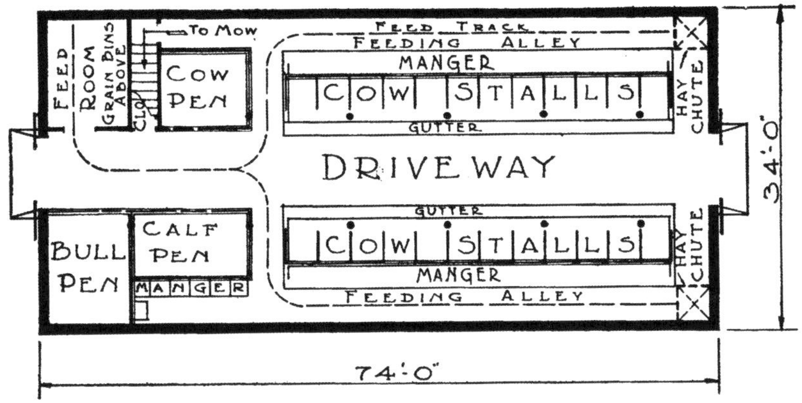

Floor Plan No. 7, Suitable for Barns Nos. 2054, 2060, 2061, 2067.

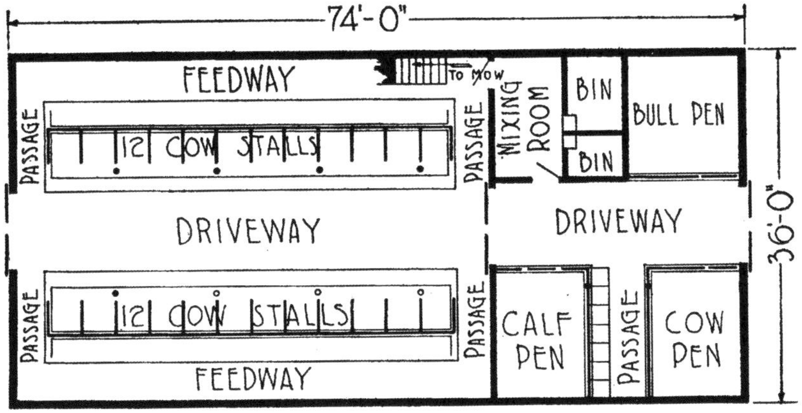

Floor Plan No. 8, Suitable for Barns Nos. 2054, 2060, 2061, 2067.

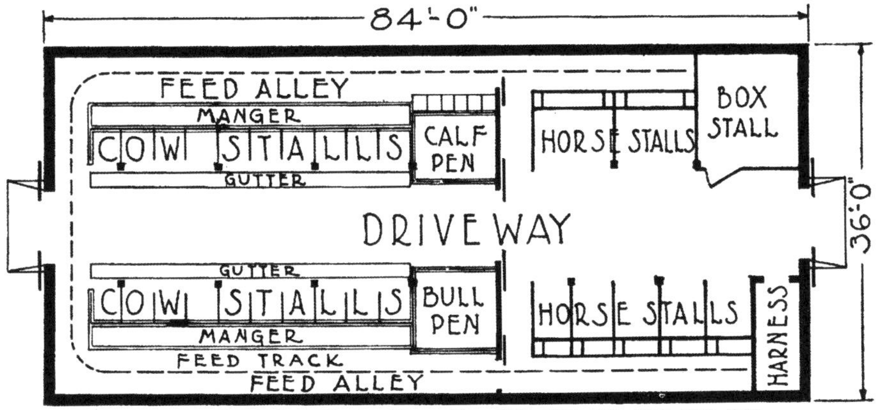

Floor Plan No. 9, Suitable for Barns Nos. 2055, 2057, 2058, 2059, 2063, 2067.

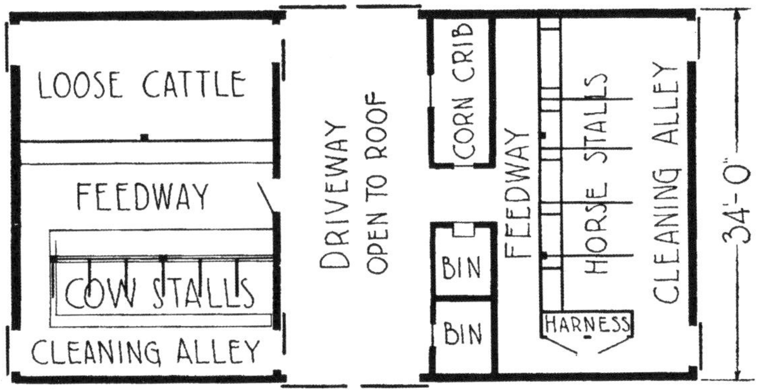

Floor Plan No. 10, Suitable for Barns Nos. 2064, 2068.

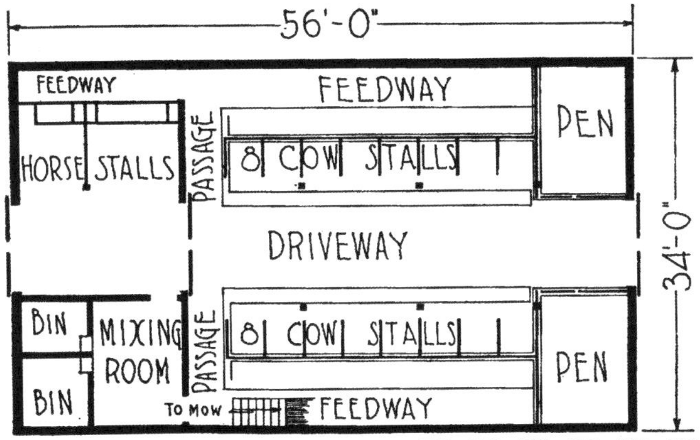

Floor Plan No. 11, Suitable for Barns Nos. 2054, 2055, 2057, 2058, 2059, 2060, 2061, 2063, 2067.

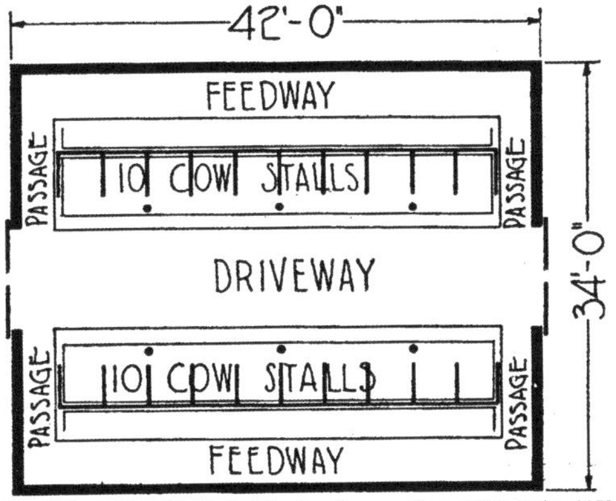

Floor Plan No. 12, Suitable for Barns Nos. 2055, 2057, 2058, 2059, 2063, 2067.

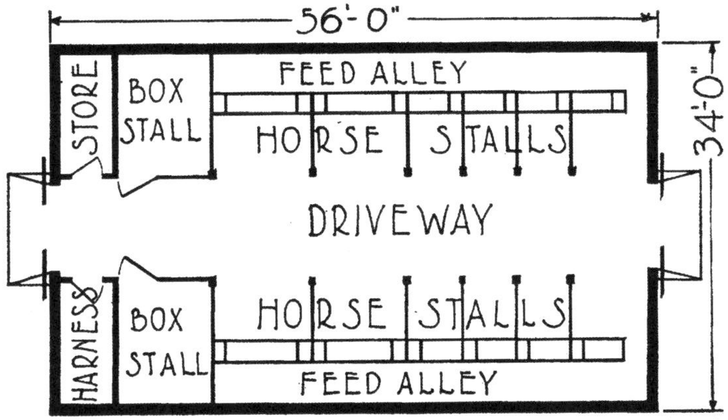

Floor Plan No. 16, Suitable for Barns Nos. 2054, 2055, 2057, 2058, 2059, 2060, 2061, 2063, 2067.

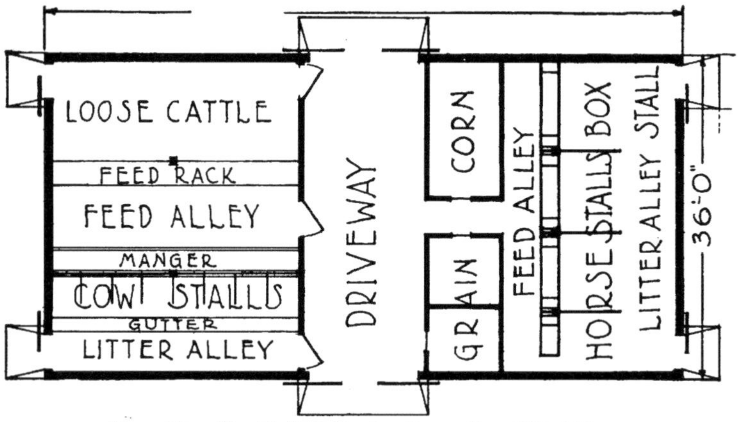

Floor Plan No. 14, Suitable for Barns Nos. 2064, 2068.

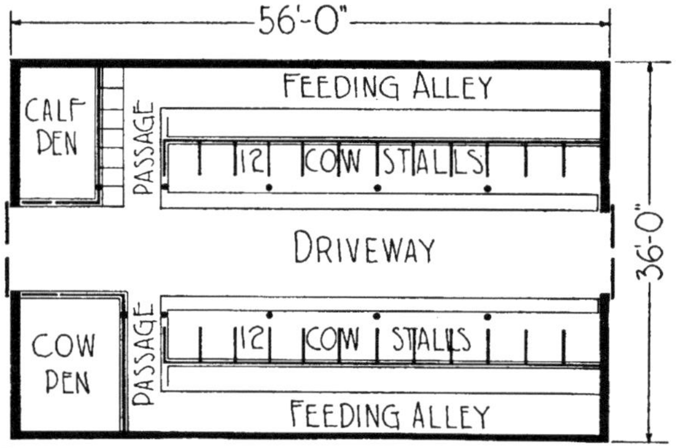

Floor Plan No. 15, Suitable for Barns Nos. 2054, 2055, 2057, 2058, 2059, 2060, 2061, 2063, 2067.

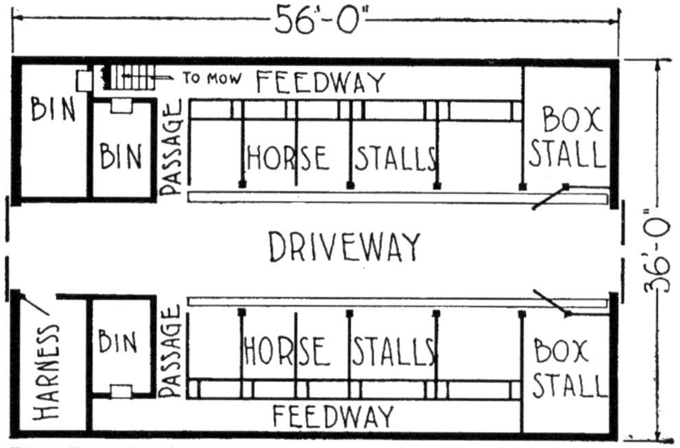

Floor Plan No. 13, Suitable for Barns Nos. 2054, 2055, 2057, 2058, 2059, 2060, 2061, 2063, 2067.

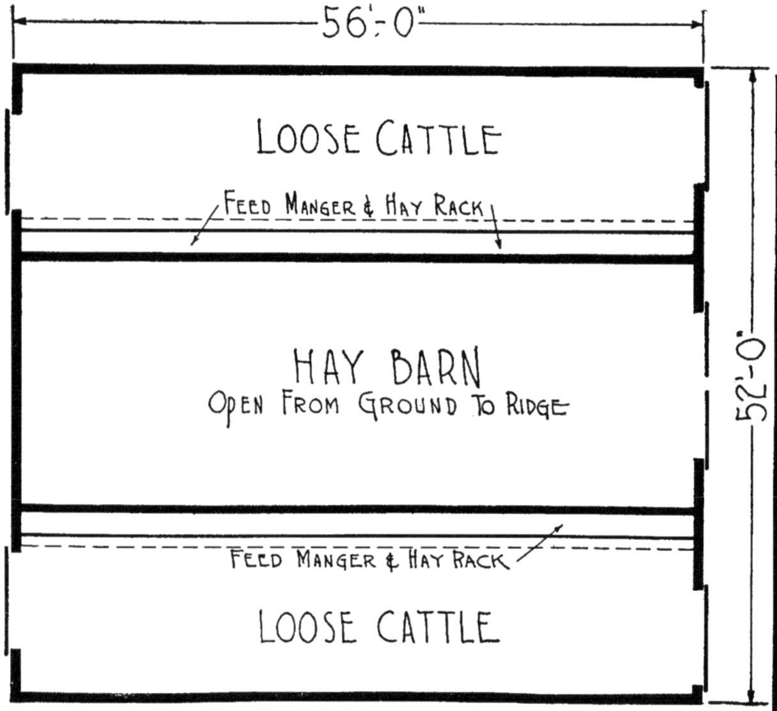

Floor Plan No. 18, Suitable for Barns Nos. 2065, 2066.

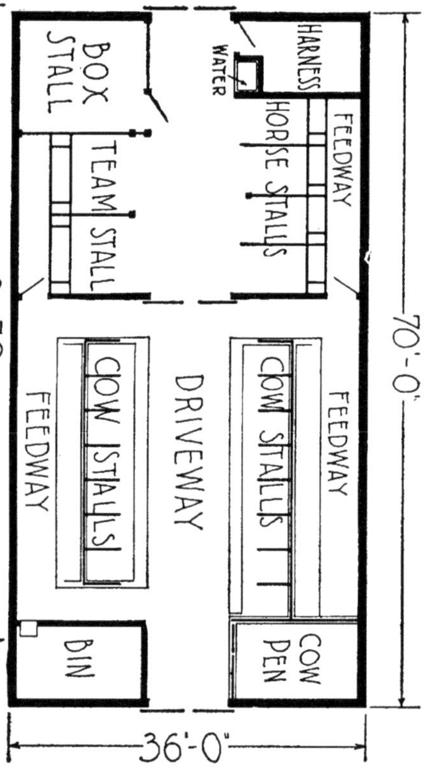

Floor Plan No. 17, Suitable for Barns Nos. 2055, 2057, 2058, 2059, 2062, 2063, 2067.

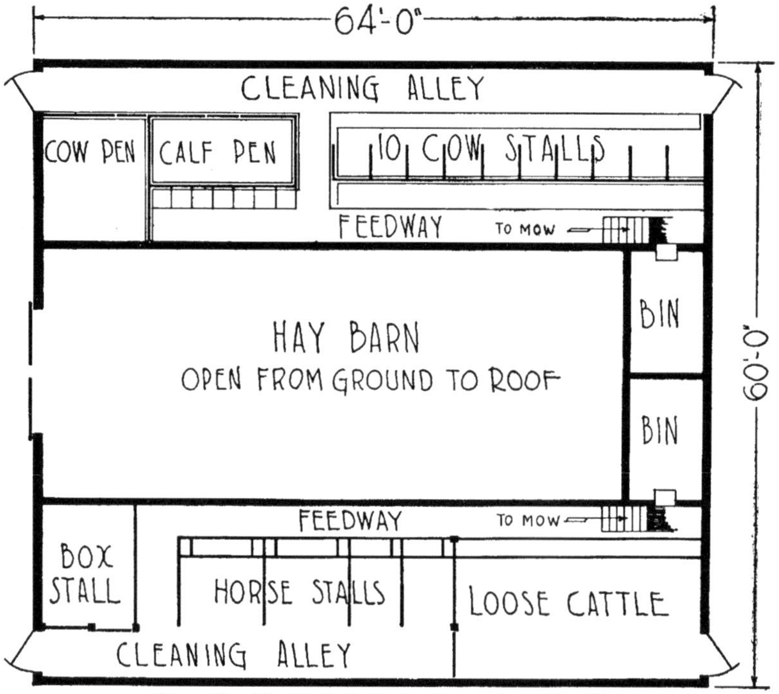

Floor Plan No. 19, Suitable for Barns Nos. 2065, 2066.

Build your barn now and save money. Prices may be much higher a few months from now. If you cannot build at once, be sure to send in your order and secure the material at present prices.

The floor plans shown on these pages fit exactly the particular barns they call for. With slight changes they can be made to fit nearly any other barn in this catalog. Select one of these plans if you possibly can.

Barn Doors

Double Sliding Barn Doors and Driveway Doors.

Catalog No.	Size	Thickness, Inches	Price, per Pair
56A5071	8 ft. 2 in. by 8 ft.	1⅝	$15.50
56A5071	8 ft. 2 in. by 8 ft.	2⅜	22.00
56A5071	12 ft. 4 in. by 12 ft.	1⅝	30.00

Frame and Hardware not included in the price.

Barn doors for the ground floor of Dairy Barns are furnished 2⅜ inches thick; all other barn doors are 1⅝ inches thick. All doors are "ready made," that is, **"already cut"** and **nailed** together, making doors superior to those made by hand carpentry. This feature is especially appreciated by our customers and saves much time and labor.

Our Driveway Doors are of the same solid construction as our other barn doors with this addition: They are supplied with or without a small built-in door in one of the wings. This built-in door permits of personal access and saves time and trouble of opening the large driveway doors. A strong hinge and hasp secure the large wings of the doors on the inside. Our barn door latch, to which a padlock may be attached, locks the built-in door and secures you against trespassers. Driveway doors are furnished in single thickness only.

Wherever conditions permit it, the barn doors in this catalog are suspended on Roll Rite Adjustable Roller Bearing Trolley Barn Door Hangers. This is the best type of barn door hangers we know of and the highest priced hangers we handle. Full description will be found on the hardware page. (See page 54.)

Double Driveway Doors With Door Cut in One of the Door Panels, as Illustrated.

Catalog No.	Size	Thickness, Inches	Price, per Pair
56A5059	12 ft. 4 in. by 12 ft.	1⅝	$35.81

Size of door in panel, 2 feet 6½ in. by 5 ft. 4½ in.
Frame and Hardware not included in the price.

Single Sliding Barn Doors.

Catalog No.	Size	Thickness, Inches	Price, Each
56A5061	4 ft. 1 in. by 8 ft.	1⅝	$ 7.50
56A5061	4 ft. 1 in. by 8 ft.	2⅜	9.89
56A5061	6 ft. 2 in. by 12 ft.	1⅝	15.76

Frame and Hardware not included in the price.

Single Sliding Barn Doors With Door in Panel, as Illustrated.

Catalog No.	Size	Thickness, Inches	Price, Each
56A5062	6 ft. 2 in. by 12 ft.	1⅝	$20.56

Size of small door in panel, 2 feet 6½ inches by 5 feet 4½ inches.
Frame and Hardware not included in the price.

Dutch Doors, as Illustrated.

Catalog No.	Size	Thickness, Inches	Price
56A5060	3 ft. 9 in. by 7 ft. 10½ in.	2⅜	$12.98
56A5060	3 ft. 9 in. by 7 ft. 10½ in.	1⅝	9.82

Frame and Hardware not included in the price.

Our Dutch Door is not generally specified for our barns. We know from experience that some builders prefer one or more additional doors. These doors serve the purpose. The Dutch Door is the ideal hinged door. In design it corresponds with the other doors. It has an upper and a lower wing. The lower wing is provided with staple and hook and the upper with a barn door latch for closing. When upper wing is locked the whole door is locked. A further advantage of this door is that the upper wing can be opened independent of the lower wing. The convenience of this arrangement will be easily understood. This door can be furnished in single or double thickness.

SLIDING GABLE HAY DOORS, 9 FEET 4 INCHES WIDE BY 12 FEET 8 INCHES HIGH. SINGLE THICKNESS ONLY.

56A5053—Price, per pair of doors........................**$32.59**
Hardware not included. Smaller sizes quoted on request.

Barn Doors *and* Windows

We have two kinds of Haymow Doors. The down sliding type has two wings, each wing hung on Roll Rite Adjustable Roller Bearing Barn Door Hangers, the highest priced hangers we handle. Each half of the door is counterbalanced by weights, to be made of concrete slab. In the slab is embedded a strong eyebolt to which is attached a cable running over pulleys and fixed at the top of the door. The counterweight should not be heavier than about half the weight of the door. All hardware is furnished and included in price when specified for our barns, but not with the doors shown on this page. Concrete slab or other counterweight to be furnished by builder.

The other type of hay door is known as the single leaf **drop hay door.** This door is of similar appearance as the down sliding hay doors. When specified for our barns, it is provided with three strong strap hinges on the outside and with a ring fastened to the inside of the door, by means of which, door may be closed or opened, by engaging the hay carrier track.

SINGLE LEAF DROP HAY DOOR, 9 FEET 4 INCHES WIDE, 12 FEET 8 INCHES HIGH. SINGLE THICKNESS ONLY.

56A5055—Price **$35.37**
Hardware, hay carrier track, etc., not included. Smaller sizes quoted on request.
Owing to its large size, this door will be shipped in two pieces, which can be easily nailed together on the job.

Ventilation Shields

CYPRESS VENTILATION SHIELD FOR DAIRY SASH OR GENERAL BARN SASH, "ALREADY CUT" AND FITTED.

56A5079—Price, per set, for one window **86c**

DAIRY SASH, 1⅜ IN. THICK; 6 LIGHTS, 8 IN. BY 12 IN. EACH. OPENING SIZE, 1 FT. 10½ IN. BY 3 FT. 6 IN.
63A5080—Price, each.........**$1.75**

For our Dairy Barns we use long and narrow sash exclusively. We make the sash special, as the ordinary market strengths of only 1⅛ inch do not satisfy our demands for good quality. Our Dairy Barn Sash is 1⅜ inches thick and consists of 6 lights, 8x12 inches; opening size, 20½ inches wide by 42 inches long. For most of the general purpose barns we use a special sash, which is also 1⅜ inches thick, consisting of 9 lights, each 9x12 inches; opening size, 32 inches wide and 42 inches high. We illustrate and price both kinds on this page for the benefit of customers who may wish to install them in old buildings.

GENERAL BARN SASH, 1⅜ INCHES THICK; 9 LIGHTS, 9 INCHES BY 12 INCHES EACH; OPENING SIZE, 2 FEET 8 INCHES BY 3 FEET 6 INCHES.
63A5081—Price, each.............**$2.18**

All Dairy Barn Sash and General Barn Sash wherever specified for the ground floor of the barn are provided with ventilation shields. All sash are also provided with spring bolts to keep the windows closed. When ventilation seems desirable, the spring bolts will hold the window in the ventilation shields. The illustration explains the simple working of this arrangement. For the convenience of those of our customers who would like to purchase additional doors, sash, etc., we have priced several of these items. The prices, however, do not include any hardware. The latter may be selected from page 54 of this catalog or from our special Hardware Catalog which will be mailed free on request.

SPECIAL BARN SASH, 1⅜ IN. THICK; 4 LIGHTS, 9 IN. BY 14 IN. EACH; OPENING SIZE, 1 FT. 10½ IN. WIDE BY 2 FT. 9½ IN. HIGH.
63A5082—Price, each..................**98c**

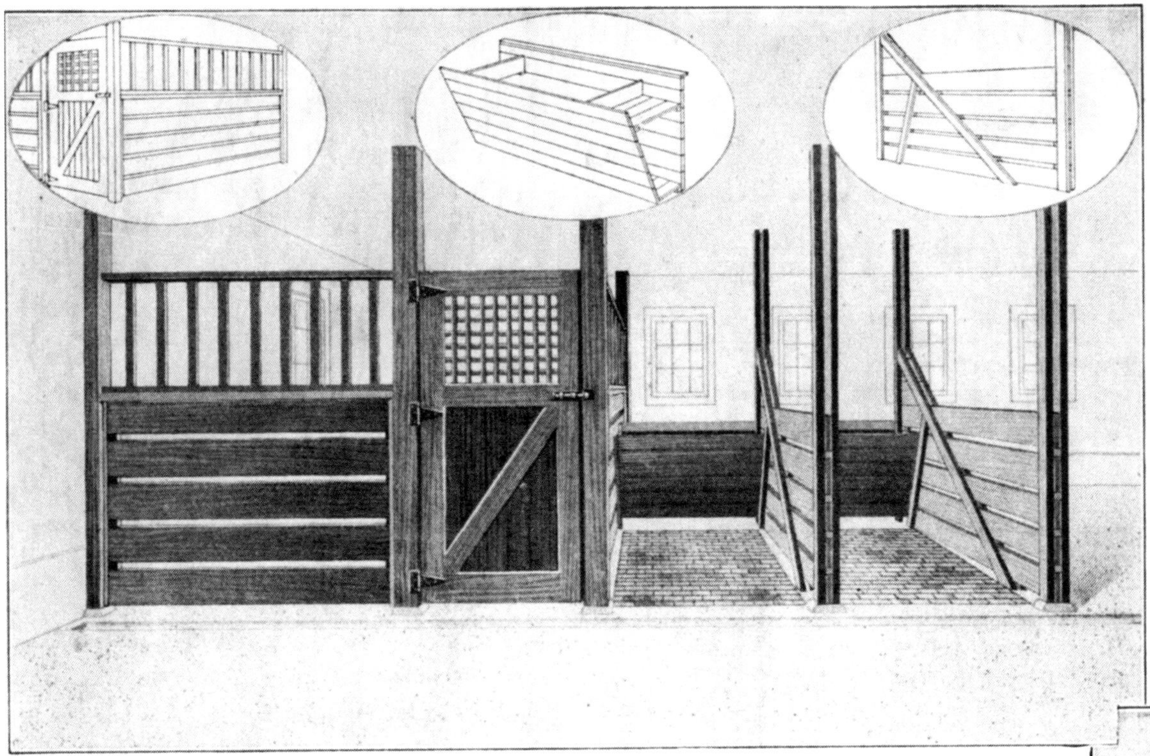

Prices for Material "Already Cut" and Fitted

BOX STALLS AND PARTS.

56A5072—Doors, 7 feet 4 inches high, 3 feet 8 inches wide, **"already made."** Price, each **$9.05**

56A5116—Partitions for Box Stalls. Price, per lineal foot .... **.74**

56A5117—Posts for Box Stalls, 6x6 inches, solid, 9 feet high. Price, each **1.61**

56A5118—Complete Box Stall, including one door, **"already made,"** 16 feet of partition and two solid posts, sizes as shown at left, including all hardware. Price **$25.00**

HORSE STALL PARTITIONS AND PARTS.

56A5119—Partition between horse stalls, 9 feet long, with built-up posts, 9 feet high. Price, each **$8.05**

56A5120—Mangers between horse stall partitions. Price, per lineal foot **1.01**

56A5121—One complete partition and one manger 5 feet long, including one grain feed box and nails. Price, complete **$12.97**

56A5122—One complete partition and one manger 9 feet long, including two grain feed boxes and nails. Price, complete **$16.90**

Box Stalls, Stall Partitions and Cattle Feed Racks

THE ECONOMY and the good appearance of our box stalls, stall partitions and cattle feed racks will appeal to you. They are strongly built, serviceable and reasonably priced.

BOX STALLS are 7 feet 4 inches high. They are made of 2-inch planks planned to be spaced 2 inches apart to secure good air circulation and mounted by a grille of 2x2-inch material capped by a 2x4-inch rail. The door is **"already made"; that is, "already cut"** and **nailed,** making a door superior to one produced by hand carpentry. It is 7 feet 4 inches high and 3 feet 8 inches wide and hung on three heavy T hinges. A solid bolt for fastening the door is provided. Note the lattice panel. It is neat and sanitary. Posts are 6x6-inch, solid, and 9 feet high. Because requirements of size vary, we have quoted posts, partitions and door separately, the price of the door including the hardware. **In ordering please specify the number of posts wanted, the length of partitions, the number of each length and the number of doors.** We shall then send the box stalls **"already cut,"** all you have to do is to **nail** them together.

STALL PARTITIONS. They are made of heavy 2-inch planks, spaced 2 inches apart. The planks are made to fit between the posts. These posts are built-up of two uprights, 2x6 inches, 9 feet high, blocked apart. In addition, two braces are provided on each side to make partitions solid and strong. Hay manger and feed box have removable bottoms to facilitate cleaning. Hay manger does not extend down to floor line. Enough room is left for a cat to get under. This prevents rats, mice and other vermin from nesting. As in the case of box stalls, we have priced posts, partitions and mangers separately to enable you to order just what you need. We have also quoted a complete partition, consisting of two built-up posts, the partition, a 5-foot manger with feed box, and a 9-foot manger with two feed boxes for team stalls.

Our CATTLE FEED RACKS will be found very serviceable. They may be used for our feeder barns. They permit of being fitted in nearly every barn where cattle are fed. The illustration shows how they are fitted against the barn wall. The light shaded parts are not included in the price. Order per lineal foot.

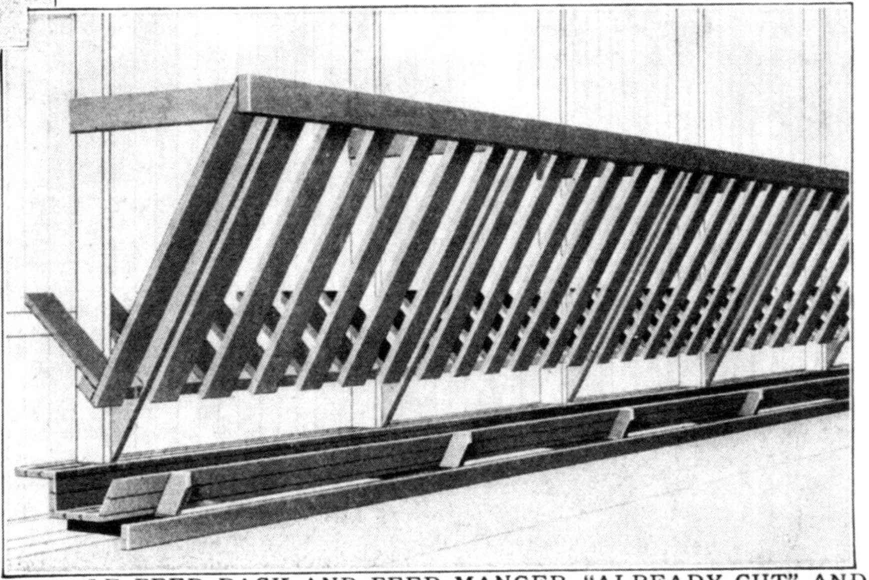

CATTLE FEED RACK AND FEED MANGER "ALREADY CUT" AND FITTED.

56A5109—Feed Rack only. Price, per lineal foot **$0.61**
56A5110—Manger only. Price, per lineal foot .... **.53**
56A5111—Complete Feed Rack and Manger. Price, per lineal foot **1.14**

"Already Cut" Hog House Pen Partitions and Shipping Crates

56A5114—Side Panel Hog House Pen Partition, "Already Cut" and Fitted.
Price, per Panel, Including Hardware.......... $5.44
Price, per Panel, Without Hardware.......... 2.85

OUR interchangeable hog pen partitions are the most practical yet designed.

THE SIDE PANELS are built like a gate to fit a pen 8 feet deep. They are 3 feet high and composed of six 1x6-inch boards, the upper two spaced about 2 inches apart. They are reinforced on one side by three vertical and two diagonal battens, as shown in the illustration at the left. These gates are hung on "No-Sag" Adjustable Gate Hinges. The lower hinge is quickly adjustable by ratchet action to take up the sag, so the gate always swings clear of the ground. The gate may also be raised to allow small animals to pass under it. This feature is of great convenience and importance. Two barrel bolts keep the pen partition securely in place. If you want to remove one of these gate partitions, push back the bolts and lift gate off. This facilitates cleaning of pens. You may also increase or reduce size of pen at will. In fact, this gate partition is of such general practical advantage that we recommend no other.

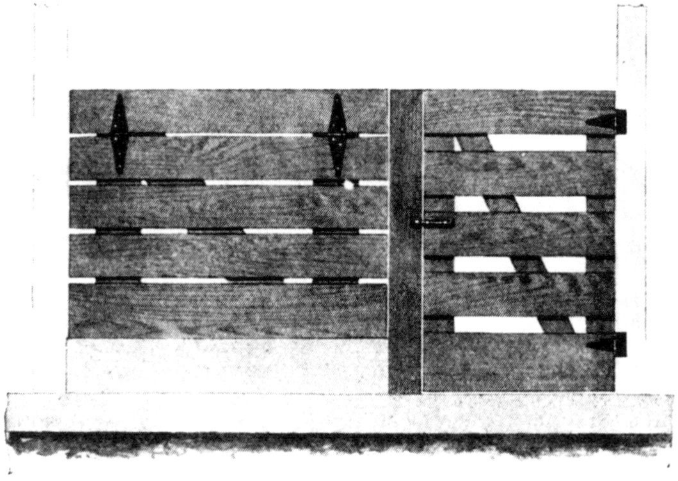

56A5115—Front Panel Hog House Pen Partition, "Already Cut" and Fitted.
Price, per Front Panel as Shown, Including Hardware.......... $5.69

THE FRONT PANELS consist of one door and a swinging feed panel over the trough. Door is hung on solid T hinges and secured by an outside bolt. Swinging feed panels hang on solid strap hinges. They open to the inside and close against the outer edge of trough.

"ALREADY CUT" CRATES. The crates shown here are "already cut" and bundled. They may be **nailed** together in a few minutes. Hang your saw on the hook and never touch it. These crates are the result of practical experience. The bottom is to be built on two runners, on which the flooring is nailed. This method of construction permits sliding the crate into cars without "racking" or breaking the crate.

Hog Crate 56A5112.

Crates are made of 1x6-inch and 1x4-inch material, trimmed down to standard sizes. All the lumber used for these crates is of good, sound quality. Two points guided us in the construction of these crates—strongest possible construction and lightest shipping weight. You will not pay for "dead" weight freight when using our crate. The price of our crates is so low, you can hardly afford to waste time and money by buying and sawing the lumber. Remember, every crate is "already cut" and bundled, nothing left for you to do but nail the pieces together. The necessary nails are included in the price.

Prices of Crates.

Catalog No.	Length, Inches	Inside Width, Inches	Height, Inches	Price, Cut and Bundled, Each
56A5112	36	12	24	**$1.02**
56A5112	48	16	30	**1.66**
56A5112	54	18	34	**1.87**
56A5113	64	25¼	36	**2.53**
56A5113	72	26	40	**2.70**

NOTE—Hog Crates 56A5112, 16 and 18 inches wide, have four side and four end bars each; that is, **ONE MORE** than shown in illustration.

Hog Crate 56A5113.

Lengthen the Life of Your Buildings by Keeping Them Always Painted With Seroco Paint

Seroco Mineral Barn Paint is an excellent preservative. The minerals are ground and reground to unusual fineness by modern up to date machinery and the oils are thoroughly mixed with the mineral pigment in large batches, thus not only securing exceptional spreading qualities but insuring uniformity of color. By giving you the proper quantity of Seroco Mineral Barn Paint with your barn we are giving you a paint that will preserve it and lengthen its life.

Although we specify Oxide Red Seroco Mineral Barn Paint for the body of all barns shown in this catalog, you may have your choice of the following colors if you prefer:

Mineral Barn Paint

30A800—Oxide Red
30A820—Dark Gray
30A830—Yellow
30A850—Maroon

We furnish enough Barn Paint with every barn for two good coats, but you will want to paint your other barns and outbuildings. Order the additional paint and include the money in your order for your barn.

Prices of above colors.	Per Gal.
50-gal. barrel..(Shpg. wt., 700 lbs.)	**$1.60**
25-gal. half bbl.(Shpg. wt., 350 lbs.)	**1.62**
5-gallon kit..(Shpg. wt., 70 lbs.)	**1.66**
3-gallon can.(Shpg. wt., 45 lbs.)	**1.67**
1-gallon can.(Shpg. wt., 15 lbs.)	**1.70**

Beautify ALL Your Buildings

Seroco Paint will make your farm buildings beautiful and attractive. It you want to make the old home new, give it a coat or two of Seroco House Paint.

For the trim of your barns we furnish 30A243, Outside White Seroco House Paint. However, if you prefer, you may have your choice of any of the following colors:

Seroco House Paint

30A215—Cream
30A207—Lemont Stone
30A259—Light Gray
30A235—Light Stone
30A218—Buff
30A244—Medium Green

Prices of Seroco House Paint.

50-gal. bbl., per gal..(Shpg. wt., 750 lbs.)	**$2.38**
25-gal. ½ bbl., per gal.(Shpg. wt., 375 lbs.)	**2.40**
5-gal. can, per gal...(Shpg. wt., 75 lbs.)	**2.44**
3-gal. can, per gal...(Shpg. wt., 45 lbs.)	**2.45**
1-gallon can ....(Shpg. wt., 15 lbs.)	**2.48**
½-gallon can........(Shpg. wt., 10 lbs.)	**1.26**
1-quart can ....(Shpg. wt., 5 lbs.)	**.66**

For actual colors of these paints, write for our General Catalog. We will send it postpaid on request.

MAJESTIC VENTILATORS

For Barns and Stables

Our Majestic Ventilators are scientifically designed and constructed. They have been planned after the best and most approved models on the market.

The size of ventilator required is governed by the number of stock in the building. If you have more than twenty head under one roof you should have two or more ventilators of the larger size.

Each ventilator has a continuous draft arrangement to draw out the foul air and replace it with fresh air, without causing a draft.

Any one of these ventilators will go a long way in making any outbuilding attractive, but its main service consists in keeping the building perfectly ventilated under all weather conditions. Leading agricultural authorities agree that ventilation is absolutely necessary in order to get the best results from stock of any kind. Animals will thrive better in a building continually supplied with pure, fresh air. Considering the low price and the high quality of our ventilators, no farmer can afford to be without one.

Extra heavy 22, 24 and 26-gauge galvanized steel is used in the manufacture of our Majestic Ventilators, according to size. Nothing thinner than 26-gauge galvanized steel is permitted. An unusually thick stock.

Each ventilator is equipped with four stay rods which hold it securely in place on the roof. A wire screen is provided to keep the birds out. When properly secured to the roof, our ventilators, owing to their weight and scientific attachment, will resist unusually severe storms.

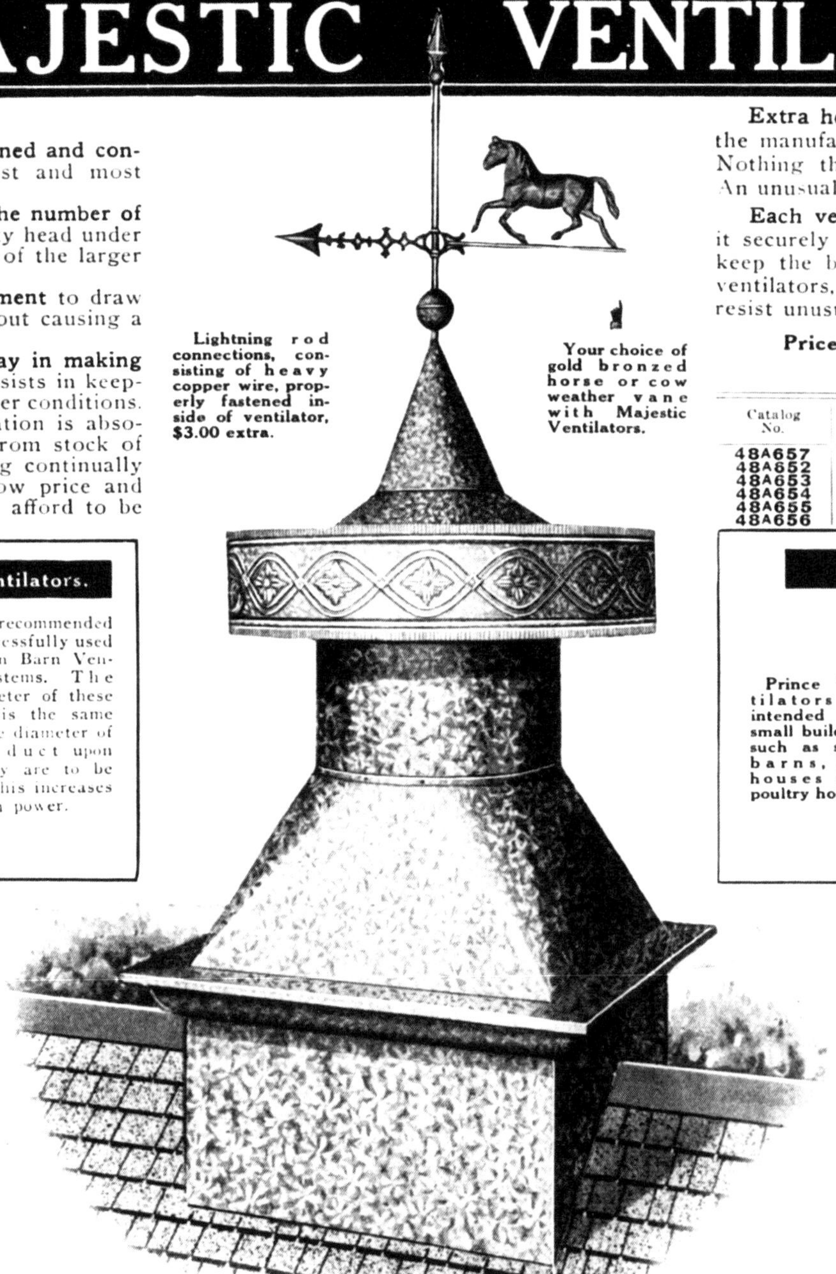

Lightning rod connections, consisting of heavy copper wire, properly fastened inside of ventilator, $3.00 extra.

Your choice of gold bronzed horse or cow weather vane with Majestic Ventilators.

Prices of Majestic Ventilators for Large Buildings.

Shipped From OHIO Factory.

Catalog No.	Size of Flue, Inches	Size of Base, Inches	Height	Wt., Lbs.	No. of Cattle	Price
48A657	18	27	7 ft. 6 in.	110	4	$22.15
48A652	20	29	8 ft. 8 in.	125	6	23.50
48A653	24	36	9 ft. 5 in.	160	8	29.95
48A654	28	40	10 ft. 4 in.	195	12	33.35
48A655	30	44	10 ft. 6 in.	205	14	36.80
48A656	36	55	11 ft. 3 in.	265	20	44.55

Leader Suction Ventilators.

Highly recommended for, and successfully used with Modern Barn Ventilating Systems. The inside diameter of these ventilators is the same as the inside diameter of the flue or duct upon which they are to be mounted. This increases their suction power.

Prices of Leader Suction Ventilators for Large Buildings.

Shipped From OHIO Factory.

Catalog No.	Size of Flue, Inches	Height, Inches	Weight, Pounds	Price
48A647	18	41	60	$10.75
48A648	20	60	68	12.60
48A649	24	68	80	15.25

Prince Ventilators for Small Buildings.

Prince Ventilators are intended for small buildings such as small barns, hog houses and poultry houses.

Prices of Prince Ventilators for Small Buildings.

Shipped From OHIO Factory.

Catalog No.	Size of Flue, Inches	Size of Base, Inches	Height	Wt., Lbs.	Price
48A650	16	22	3 ft. 5 in.	45	$ 9.30
48A651	18	24	3 ft. 6 in.	60	11.00

$120^{00} BUYS AN 8x20-FOOT SELECT YELLOW PINE COLUMBIA SILO

The Columbia Silo is guaranteed to give perfect satisfaction and service. It is shipped to the purchaser with the understanding that if it is not exactly as we represent it, the entire silo, or any part of it, can be returned to us at our expense and the purchase price and all transportation charges will be promptly returned.

Columbia Silos are furnished with full length Clear Fir Staves, this being our highest grade, and full length Select Fir Staves, our next grade below Clear. We also furnish Columbia Silos made with Select and Better Fir Spliced Staves and Select Yellow Pine Spliced Staves, same being joined together with galvanized steel splines.

CLEAR GRADE is our highest grade and is practically clear of all defects, as its name implies, furnished only in Fir.

SELECT GRADE Fir or Yellow Pine contains minor defects, which, however, will not injure its use for a good, tight, substantial silo.

Spliced Staves are furnished in SELECT and BETTER GRADE Fir and **Select Yellow Pine.** Two pieces of these staves are joined together with a galvanized steel spline to make the required length.

All of our staves are worked from 2-inch stock planed on both sides, edges worked with tongue and groove.

Do not let our low prices bring any question to your mind concerning the quality of our Silos. We never sacrifice quality for price.

DOORS AND FRAMES—Our door frame is one of the strongest on the market, made of the same quality of material as our staves. The cross braces are made of 2x4s bolted to the uprights. The frame is further reinforced by a ¼x2½-inch bar of steel, which gives great strength to the door frame. The door is so arranged that the inside is practically smooth, allowing the silage to settle freely, as it should.

OUR SILOS ARE TIGHT—The doors are interchangeable and are drawn to place by means of a heavy tail nut and bolt on each side which clamps a heavy piece of steel against the outside of the frame.

FELT GASKETS—Are provided to be tacked on the doors.

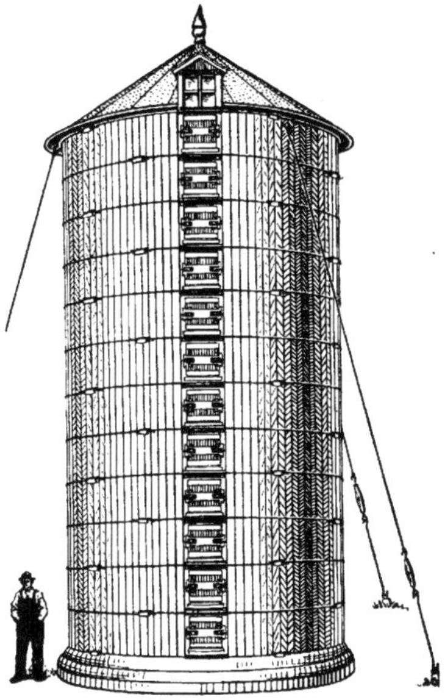

HOOPS AND LUGS—Hoops are made of ⁹⁄₁₆-inch steel in three or four sections for each hoop, depending on the diameter of the silo, and have standard machine cut threads extending 7 inches from each end, which gives 3 feet and 6 inches draw on each hoop. Strong malleable iron lugs and clean, well fitted nuts join the ends of the hoops.

SAVE ON FREIGHT—If a few men club together, ordering four or five silos, the freight will be reduced considerably by shipping in carload lots (approximately 30,000 pounds). Send us the names of your neighbors who are interested.

If you are unable to secure a sufficient number of neighbors to join you in ordering a carload, send us your own order anyway. Send it early, specifying the latest shipping date that will be convenient to you, and we may be able to include your silo with others coming near to your station.

Galvanized Steel Roofs.

26-gauge steel. This roof is self supporting, the standing seams of four thicknesses of steel serving as rafters. Filling door and ventilator included.

Prices of Galvanized Steel Roofs Without Dormer or Sash.

To Fit Silo Diameter (Outside)	56A1245 Price	Weight
10 feet	$28.80	190 pounds
12 feet	36.80	275 pounds
14 feet	46.15	355 pounds
16 feet	56.80	279 pounds
18 feet	68.80	550 pounds
20 feet	82.15	685 pounds

56A1248—Steel Dormer and Glazed Sash, with hinges and screws for dormer, extra...............**$8.55**
Shipped from factory in OHIO.

MONEY SAVERS.

Felt Gaskets for Any Silo.

Be sure to tell us the name and size of your silo and we will send you the proper quantity of good felt gaskets, ready to tack on. Do not delay ordering this until too late, as doors that are not tight cause the silage to decay.

56A1250—Felt Gaskets. Price
For silos 24 feet or less in height................**$2.00**
For silos 24 to 30 feet high........................ **2.50**
For silos 30 to 36 feet high........................ **3.00**

The above prices include sufficient quantity for a complete silo of size given. Shipping weight of any of above gaskets, each, about 1 pound.

Silo Door Hinges.

Allow the Door to Swing Back Into Silo.
Price, **85c** per Set Extra for Each Door.

When ordering this patent door hinge for Columbia Silos be sure to order **56A1246.**

SAVED $50.00—LUMBER A1.

Sears, Roebuck and Co., Chicago, Ill. Chagrin Falls, Ohio.
Gentlemen:—I am well pleased with the silo I purchased from you and have saved about $50.00 on the order over any price I was able to get elsewhere. The lumber is A1. Very truly yours, D. D. GORE.

Prices of Yellow Pine and Fir. Columbia Silos Loaded on Cars at Factory.

Diameter, Feet	Height, Feet	Number of Hoops	Estimated Capacity, Tons	Number of Cattle	Approximate Weight, Without Roof, Pounds	Full Length Clear Fir Staves Without Roof, but With Anchors, Cables, Turnbuckles and Hoops 56A1241	Full Length Select Fir Staves Without Roof, but With Anchors, Cables, Turnbuckles and Hoops 56A1240	Spliced Select and Better Fir Staves Without Roof, but With Anchors, Cables, Turnbuckles and Hoops 56A1242	Spliced Select Yellow Pine Staves Without Roof, but With Anchors, Cables, Turnbuckles and Hoops 56A1244
8	20	9	20	6	3,200	$160.00	$153.00	$149.00	$120.00
8	24	11	25	7	3,850	219.50	212.00	204.50	160.50
10	20	9	30	8	4,400	216.00	207.50	199.00	161.00
10	24	10	36	11	5,250	258.00	248.00	238.50	192.00
10	28	13	44	13	6,100	299.50	288.50	277.00	223.00
10	30	13	48	14	6,525	322.50	313.00	296.50	238.00
10	32	14	53	15	6,950	344.00	331.50	315.50	254.00
12	20	9	45	12	5,200	250.00	240.00	230.00	182.50
12	22	10	50	14	5,725	271.50	261.50	250.00	199.00
12	24	11	55	16	6,250	296.00	284.50	272.00	218.00
12	26	12	60	16	6,775	321.00	306.00	294.50	234.00
12	28	12	68	18	7,300	346.00	332.00	319.50	254.50
12	30	13	76	22	7,800	379.00	365.00	350.50	269.00
12	32	14	86	25	8,300	392.00	377.00	361.00	287.00
12	34	15	94	27	8,800	417.00	400.50	382.50	305.00
14	20	9	60	16	6,050	281.00	269.50	257.50	203.00
14	22	10	66	18	6,650	309.00	295.50	282.00	223.00
14	24	11	73	21	7,250	337.00	323.00	308.00	243.00
14	26	12	80	23	7,850	365.00	348.50	341.00	263.50
14	28	12	90	26	8,450	390.00	373.50	357.00	284.50
14	30	13	100	28	9,000	432.00	402.50	396.50	300.00
14	32	14	110	32	9,600	452.50	428.00	407.50	321.50
14	34	15	120	34	10,200	472.50	455.00	433.50	344.00
16	24	11	95	27	8,350	376.50	361.00	344.00	503.00
16	28	12	111	33	9,650	440.50	422.00	402.50	315.50
16	30	13	120	35	10,200	484.50	463.50	444.00	334.00
16	32	14	130	38	10,850	499.50	480.00	455.50	356.50
16	34	15	142	40	11,500	530.50	510.50	483.50	379.00
18	28	12	140	40	10,200	481.50	463.00	439.50	342.00
18	30	13	150	44	10,800	529.50	510.50	484.00	363.00
18	32	14	162	46	11,500	545.50	522.00	496.00	387.00
18	34	15	180	50	12,200	579.50	563.00	528.00	411.00
20	28	12	170	49	11,200	531.50	509.00	483.50	373.50
20	30	13	185	52	12,000	582.00	555.50	531.00	395.50
20	32	14	200	56	12,800	598.50	574.50	544.00	421.50
20	34	15	220	63	13,600	636.00	606.00	578.50	448.50

Prices for Complete Hoops.

With lugs, nuts, wrench and draw bar.

Diameter	56A1261 Price, Each	Weight
8 ft.	$2.35	30 lbs.
10 ft.	2.85	39 lbs.
12 ft.	3.50	48 lbs.
14 ft.	4.00	55 lbs.
16 ft.	4.50	62 lbs.
18 ft.	5.00	69 lbs.
20 ft.	5.65	78 lbs.

Yellow Pine Silos Shipped From SOUTHERN ILLINOIS; Fir Silos Shipped From Factory in MINNESOTA.

Anchor Complete.

56A1262

Height	About	Price, per Set of Three
20 to 28 ft.	100 lbs.	$6.70
28 to 34 ft.	100 lbs.	7.90

Malleable Iron Lug for Silo Hoops.

56A1264
Weight, 1 pound.
Price, each...................................**15c**

High Grade Equipment for Barns

Combination Sliding or Swinging Barn Door Latch.

30c Without Padlock

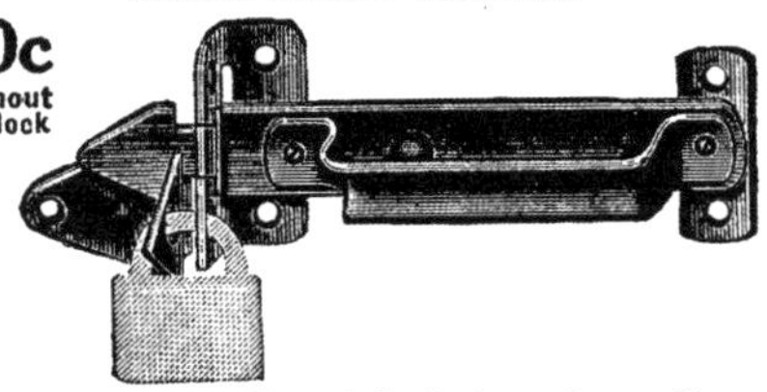

This high grade all steel latch is used on all our barns and many other buildings. Can be used on right or left hand swinging or sliding doors and locked with a padlock. Sets flush with inside surface of door, permitting rolling or sliding doors to be opened their full width. Shipping weight, about 1½ pounds.

9A4233—Price, each, complete with screws for attaching, but without padlock......................**30c**

"No-Sag" Adjustable Gate Hinge

$1.48

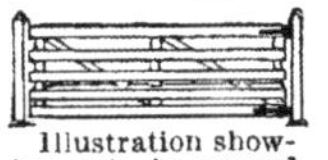

Illustration showing gate in normal position.

Illustration shows gate raised to permit small animals to pass under.

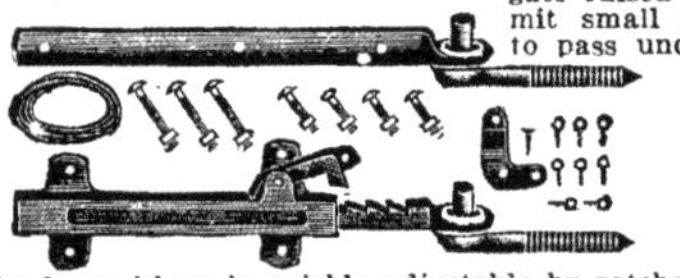

The lower hinge is quickly adjustable by ratchet action to take up the sag so gate always swings clear of the ground. Also gate may be raised to allow small animals to pass under. Furnished complete with necessary bolts, screws, etc., and directions for attaching. Shipping weight, about 10 pounds.

99A4431—Price, per set for one gate......**$1.48**

Standard Pattern Dehorning Clippers.

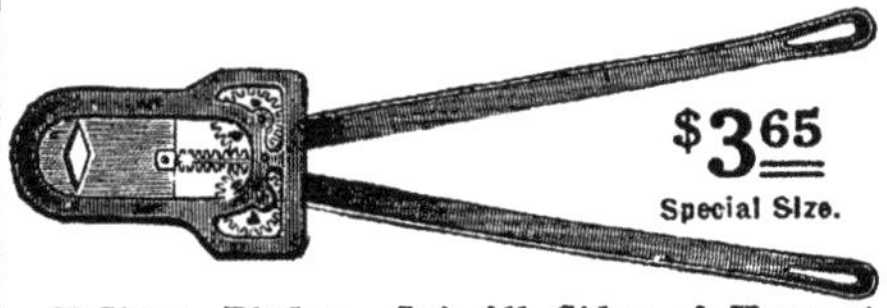

$3.65 Special Size.

V-Shape Blades. Cut All Sides of Horn at Once.

Compound levers make cutting easy. Handles and V head frame are of malleable iron. Knives are of steel, will not interlock or cut into each other.

99A5908—Special Size Dehorner. Width of opening, about 2¾ inches. Shipping weight, about 13 pounds. Price **$3.65**

Extra Knives for Special Size.

9A5942—Sliding Knife. Weight, about 12 ounces. Price **45c**

9A5941—Stationary Knife. Weight, about 8 oz. Price **35c**

99A5909—Large Size Dehorner. Width of opening, about 3¾ inches. Shipping weight, about 14 lbs. Price **$4.45**

Extra Knives for Large Size.

9A5944—Sliding Knife. Weight, about 1 pound. Price **55c**

9A5943—Stationary Knife. Weight, about 10 oz. Price **45c**

"ROLL-RITE" ADJUSTABLE ROLLER BEARING TROLLEY BARN DOOR HANGERS.

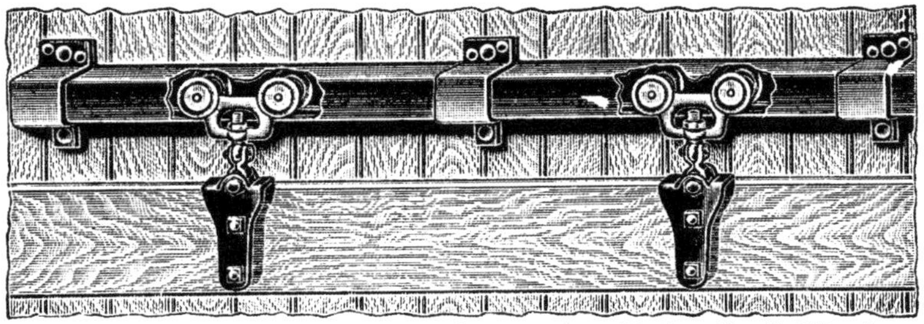

$1.40 PER SET

No binding.
No sagging.
No rusting.
Troubleproof.

The best trolley barn door hangers are the only ones to give satisfaction. For all sliding doors of any of the barns shown in this catalog, we have selected "Roll-Rite" Adjustable Roller Bearing Trolley Barn Door Hangers. There is no binding, no sagging, no rusting. It is the best type of barn door hanger made, because it has both lateral and vertical adjustment, which gives any alignment desired without the necessity of blocking up or removing the door. Heavy, rustproof, malleable iron frame. Electro galvanized steel wheels, cold rolled steel axles. Made to stand hard usage on the heaviest doors. Set consists of two hangers (one pair), one center bracket and two end brackets for track. Suitable for 1¾-inch doors. Shipping weight, about 10 pounds.

99A4468—Set complete without track.
Price, with necessary bolts and screws for attaching......................**$1.40**

99A4469—Heavy Wrought Steel Japanned Trolley Track for above hangers. Price does not include brackets. **State length.** Shipping weight, about 1¾ pounds to the foot.

Length, feet	4	5	6	8	10
Price, per length	**84c**	**$1.05**	**$1.26**	**$1.68**	**$2.10**

9A4241—Wrought Steel Japanned Center Brackets for above track. Brackets should not be more than 3 feet apart. Price includes lag screws. Weight, about 12 ounces.
Price, each......................................**14c**

PURE MANILA ROPE.

Long Fiber, Fresh New Stock, Full Size and Strength.

Guaranteed all pure manila, no jute or sisal mixed with it. High grade in every respect. For hay carriers it is unexcelled.

Manila rope is never measured exact diameter, one-third of the circumference being considered the diameter. Can furnish in one piece any length up to 1,200 feet. **State size wanted.**

NOTE—Rope for our hay carriers should not be ordered larger than 1 inch in diameter. **99A7825**

Size	Weight, per Foot	Price, per Foot	Price, per 100 Feet
¼ inch	⅜ ounce	9/10c	$0.70
⅜ inch	1 ounce	2¼c	1.85
½ inch	1½ ounces	3¼c	2.75
⅝ inch	2 ounces	5c	4.50
¾ inch	3 ounces	6½c	6.00
⅞ inch	4 ounces	8½c	7.80
1 inch	5 ounces	10c	9.50

Kenwood High Grade Hay Carrier Equipment

We list below a few items selected from our big line of hay carrier equipment for use in our barns. For our complete line refer to our big General Catalog, a copy of which will be sent to any address postpaid upon request.

Kenwood Double Angle Steel Track.

19c PER FOOT

FOR TRACK ONLY.

14c EACH

STEEL TRACK HANGING HOOK.

9A6158

7½ inches long. Wt., 1 lb...**14c**

6c EACH

RAFTER BRACKET.

9A6160

For hanging hay carrier track. Weight, about 4 ounces. Each **6c**

For our steel track carriers. Made in 5 and 10-foot lengths only, with couplings and bolts.

99A1030—Price, per foot..........(Weight, per foot, about 2 pounds)............**19c**

Eye Bolt. 42c

9A6152—18 inches long. Weight, about 3 pounds.
Price **42c**

Floor Hook.

14c

9A6150—¾ inch diameter, 7½ inches long. Weight, 1¼ pounds.
Price **14c**

Kenwood Double Swivel Hay Carriers.

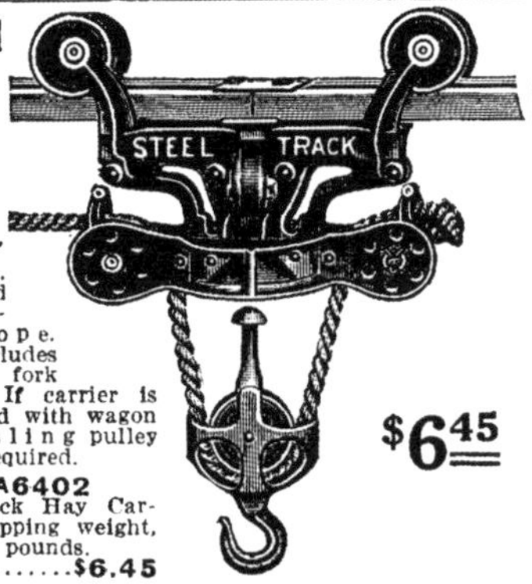

$6.45

For ¾-in., ⅞-in. and 1-in. manila rope. Price includes stop and fork pulleys. If carrier is to be used with wagon slings a sling pulley will be required.

99A6402

Steel Track Hay Carrier. Shipping weight, about 30 pounds.
Price **$6.45**

Kenwood Double Harpoon Hay Forks

99A6417—With 24-inch tines and 16 inches between points. Shipping weight, about 18 lbs.
Price **$2.20**

99A6418—Same as above. Tines, 30 inches long. Shipping weight, about 25 pounds.
Price **$2.45**

99A6419 — Alfalfa Double Harpoon Hay Fork. Extra heavy. Has tines 32 inches long, 22 inches between points. Shipping wt., about 30 lbs. Price..**$3.90**

Kenwood Adjustable Wagon Slings.

$4.70 6-Ft.

$4.35 5-Ft.

Illustration shows 5 and 6-foot slings.

$3.45 4-Ft.

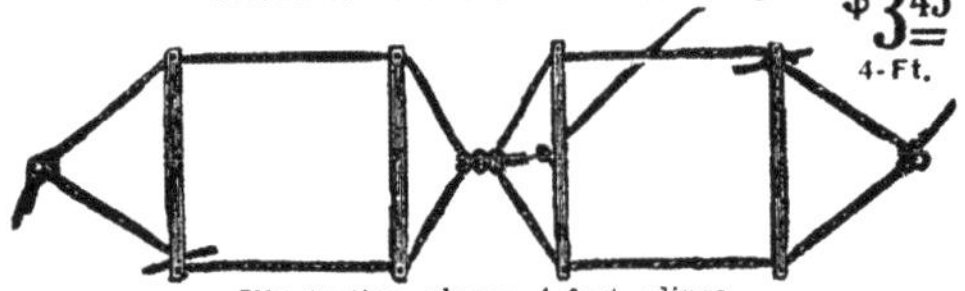

Illustration shows 4-foot slings.

Prices are for single slings, as illustrated, not for sets. Our slings are adjustable to suit any length of rack.

	Length	Shipping Weight, About	Price
99A6440	4 feet	21 lbs.	$3.45
99A6441	5 feet	26 lbs.	4.35
99A6442	6 feet	29 lbs.	4.70

Kenwood Right Angle Self Locking Sling Pulley.

$2.95

Draws up load without twisting pulley. For use with our hay carrier only. Shipping wt., about 12 pounds.

99A6448—Price **$2.95**

"Life-Time" Creosoted Wood Paving Blocks

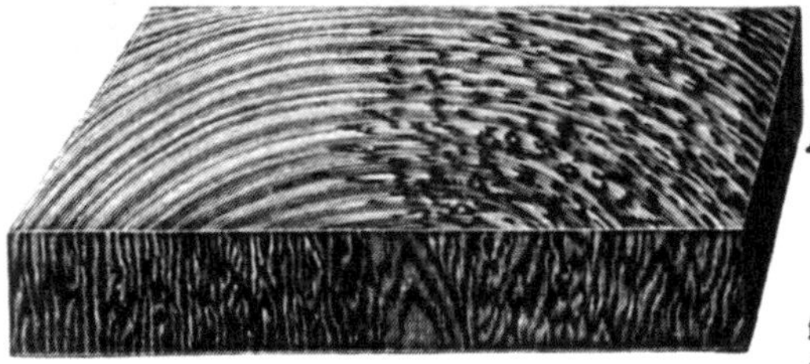

Note the Vertical Grain Which Forms the Long Wearing, Smooth Surface

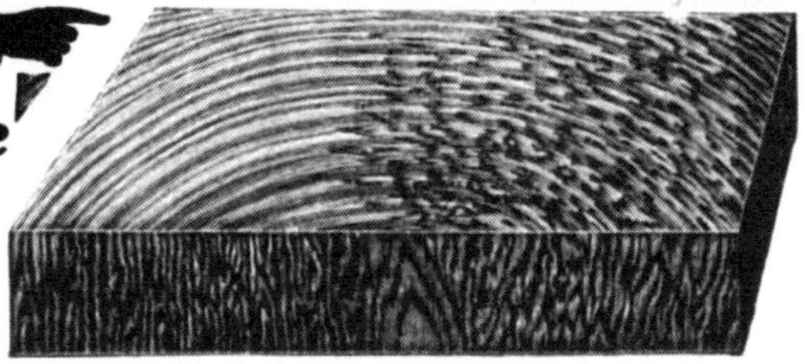

ILLUSTRATIONS SHOW SIDE TO BE PLACED UP

This durable, sanitary and economical flooring has been specially manufactured for use in dairy barns, stables, stalls, garages, driveways, fire stations, blacksmith shops and factories. By the use of these Creosoted Wood Blocks, which can be laid close, you get a smooth floor that can be easily cleaned. The blocks are made practically uniform in depth and width. They have square edges, forming a tight joint, and by the use of Mastic Pitch, described elsewhere on this page, a waterproof surface is secured.

The creosote saturation is in itself a germicide, as it contains a large percentage of carbolic acid. The oil in the block, which is forced in under heavy pressure, prevents the absorption of water to any considerable degree. A Creosoted Wood Block floor can be flushed with water as effectively as any other kind of flooring and is easily swept clean.

Our "Life-Time" Creosoted Wood Blocks will last practically as long as any building in which they are used. The same material under heavy traffic in the streets of large cities only wear out about one-eighth of an inch in six years. When you figure how much less the wear will be in a stable or barn, you can easily see that our "Life-Time" Creosoted Wood Blocks will live up to their name.

A floor of this material is much more comfortable and warm for man and beast than a concrete, cinder or earthen floor. The vertical grain insures not only long wear, but an even, smooth surface that will not splinter or crack under heavy strain.

Barn Paved With "Life-Time" Creosoted Wood Paving Blocks 2¼ Inches Thick. The Modern, Sanitary and Durable Flooring for All Outbuildings.

The process of manufacture by which the moisture in our "Life-Time" Creosoted Wood Blocks is removed and the creosote saturation driven in, calls for a treatment of several hours' duration. A pressure of about 175 pounds per square inch is maintained until the required amount of oil per cubic foot has been forced into the blocks.

"Life-Time" Creosoted Wood Blocks are easy to lay. No skilled labor is necessary. The only tool required is a hatchet to tap them into place and shave off the end of a block in case the course does not come out exactly even.

Our Creosoted Wood Blocks are much more economical, more sanitary and last longer than expensive cork blocks, which have had an enormous sale during recent years. We advise you to give our "Life-Time" Creosoted Wood Blocks a trial. After they have been once used, you will not consider any other kind of flooring for your barn, stable or any other outbuilding.

How to Order.

Determine the number of square feet in the surface to be covered. Divide this by nine, which will give the number of square yards.

Shipped from factory at Indianapolis, Ind.

Specially Prepared Mastic Pitch for Filler

This pitch is heated and poured while hot in the joints or cracks between the blocks.

56A384

300-lb. barrel......(will lay 30 square yards)......$2.85
400-lb. barrel......(will lay 40 square yards)...... 3.80
500-lb. barrel......(will lay 50 square yards)...... 4.75
600-lb. barrel......(will lay 60 square yards)...... 5.70

Pitch shipped from CINCINNATI, OHIO, and CHICAGO, ILL.

Price of "Life-Time" Creosoted Wood Paving Blocks, 2¼ Inches Thick.

Catalog No.	Width	Length	Thickness	Weight, per Square Yard	Price, per Square Yard
56A381	3 in.	8 in.	2¼ in.	80 lbs.	$3.20

Price of "Life-Time" Creosoted Wood Paving Blocks, 2½ Inches Thick.

Catalog No.	Width	Length	Thickness	Weight, per Square Yard	Price, per Square Yard
56A382	3 in.	8 in.	2½ in.	100 lbs.	$3.40

Price of "Life-Time" Creosoted Wood Paving Blocks, 3 Inches Thick.

Catalog No.	Width	Length	Thickness	Weight, per Square Yard	Price, per Square Yard
56A383	3 in.	8 in.	3 in.	120 lbs.	$3.75

No. A2035
$1,418.00
6 Rooms

BUILD YOUR HOUSE ON EASY PAYMENTS

No. A2037
$1,449.00
5 Rooms

No. A2033
$963.00
4 Rooms

No. A2030
$823.00
5 Rooms

MACHINE MADE HOUSES

Already Cut to Fit.

THE TWELVE HOUSES SHOWN ON THIS PAGE are not portable or of flimsy construction, but **double walled, double floored houses,** with 2x4 studding only 14⅜ inches apart; built warm enough for cold climates. They are twelve of our Machine Made, Ready Cut Houses. Framing lumber and mill work that go into these buildings are cut to fit by improved machinery—making perfectly true cuts and miters, far better than could be done by hand labor. YOU HAVE NO MITERING, NO MEASURING TO DO—YOU SIMPLY NAIL THEM TOGETHER.

THE MATERIAL IS GUARANTEED TO BE FIRST CLASS. We furnish Framing and Dimension Lumber, Clear Yellow Pine Flooring, Clear Red Cypress Siding and Clear Finishing Lumber, all Mill Work, such as Doors, Windows and Frames, all Building Paper, Eaves Trough, 4-Ply Fifteen-Year Guaranteed Fire-Chief Shingle Roll Roofing, all the Hardware, and Painting Material for two coats outside and inside, at the low prices quoted. We will furnish the material on EASY PAYMENTS and even ADVANCE CASH, if needed, to HELP YOU BUILD.

These are but a few of over 100 great values in our Free Book of "Honor Bilt" Modern Homes illustrated below. An entirely new book with several color pages suggesting beautiful interior and exterior treatment. Be sure to send for this book. It will show you how to cut your building costs way down and how to save ALL architects' fees. Our method is so easy and simple. Select a house from our book, send in your order, and in a short while the materials will be laid down at your station. NO WORRIES, NO WAITING for months selecting and making material to order. We do all of that for you and WE DO IT QUICKLY.

No. A2004
$945.00
4 Rooms

No. A2031
$801.00
4 Rooms

No. A2027
$853.00
5 Rooms

No. A2034
$603.00
3 Rooms

Write for This New Free Book No. 596A

No. A2002
$1,014.00
4 Rooms

No. A2028
$877.00
4 Rooms

No. A2032
$843.00
5 Rooms

No. A2029
$827.00
5 Rooms

"HONOR BILT" FARM BUILDINGS ORDER BLANK

Three Methods of Payment: 1—CASH WITH ORDER; 2—Pay After Inspecting Material; 3—EASY PAYMENTS.

IS THERE A FREIGHT AGENT at your railroad station?.....................

HOW TO SEND MONEY. The best and safest ways to send money are by Postoffice Money Order, Express Money Order, Bank Draft or Check.

DO NOT SEND COIN; it is liable to break through the envelope and be lost. If absolutely necessary to send coin BE SURE to wrap it in strong paper before putting it in the envelope, and then send by REGISTERED MAIL.

DO NOT WRITE IN THIS SPACE

SIGN FULL NAME.

DATE______________________19____

NAME______________________________________

POSTOFFICE________________________________

R. F. D. No.________ BOX No.________ STATE________________

STREET AND No.____________________________

GIVE SHIPPING POINT IF DIFFERENT FROM POSTOFFICE.

NAME______________________________________
(Give name here only when shipment is to be made to another party. If to yourself, leave space blank.)

SHIPPING POINT____________________________

COUNTY________________________STATE________________

STATE TOTAL AMOUNT OF MONEY SENT WITH THIS ORDER	DOLLARS	CENTS

CASH WITH ORDER

Sears, Roebuck and Co., Chicago, Ill.

In accordance with your guarantee of satisfaction or money returned, enclosed please find........................
(State whether Postoffice or Express Money Order, Draft, Check or Letter of Deposit.)

for $........... for your Modern Farm Building No....................described on page............of your Book
(State number and size of building wanted.)

of "HONOR BILT" Barns. The order is sent with the understanding that, in addition to the lumber, mill work and roofing your price includes nails, hardware and painting material, as well as building plans.

YOUR MONEY, INCLUDING ANY TRANSPORTATION CHARGES YOU PAID WILL BE IMMEDIATELY RETURNED TO YOU FOR ANY GOODS NOT PERFECTLY SATISFACTORY.

....
(Sign your name on this line.)

Color of paint for outside body..........................

Color of paint for outside trim..........................

Remarks...

....

....

PAY AFTER INSPECTING MATERIAL

If you would like to inspect material before sending us the money you can deposit the payment required with your bank or building and loan association and send us with your order this **letter of deposit.** This will give you a chance to inspect your material before we get a cent of our money.

LETTER OF DEPOSIT

................................19.....

Postoffice.............................. State ....

Sears, Roebuck and Co., Chicago, Ill.

Gentlemen:—

....has deposited with us the sum of $.....................
subject to your order, same to be paid to you upon delivery of the building material ordered of you, with the understanding that the goods must conform with your specifications and meet with depositor's approval. The material is to be inspected immediately upon its receipt and, if satisfactory, accepted by the depositor, who will then notify us to send you the money.

Yours truly,

....
(Name of Bank or Building and Loan Association.)

....
(Signature of Officer and Title.)

SEARS, ROEBUCK AND CO. - - CHICAGO

EASY PAYMENTS

EASY PAYMENT PLAN

If it is not convenient for you to pay cash according to the terms on the other side of this blank, we may still be able to ship you the material under our **Easy Payment Plan.**

Land Free and Clear From Debt.

If you own farm land free and clear from debt, Sears, Roebuck and Co. will ship Farm Buildings on **Easy Payment terms.** You can pay in small monthly or half yearly installments. Should it be necessary, we will allow you five years' time to pay us. Interest only 6 per cent per year, payable half yearly.

Land Not Free and Clear From Debt.

Even if you should be owing on your property, we may be able to offer you the same liberal terms.

Here Is What Would Be Required of You in Either Case:

1. Show Title to your farm land.

2. Send us with order 2½ per cent of the amount that will be due us for material. This is to cover the cost of handling the transaction, such as examining title, drawing up papers and other expenses we do not have in connection with our regular cash with order business.

3. Make small payments each month for five years on the debt. Payments may be made every three months or every six months if more convenient. You can pay as much more than your regular payment as you wish, or even take up the loan in full any time during the five years.

Here Is What Sears, Roebuck and Co. Will Do:

Ship material and give you five years to pay.

By answering the following questions, we will be able to tell you what our terms will be in your particular case.

---

1. How long have you lived in place where now located? ....

2. Do you own the land on which you intend to build? ....

3. How large is the farm?..............................

4. How much did it cost? $.................. **5.** Is it free and clear from debt?..............

6. If your property is not free from debt, please state how much you owe on it. $..................

....

7. Which of our farm buildings do you intend to put up?.................................
(Number.) (Size.) (Price.)

8. When do you expect to start building?..............................

9. What crops do you grow?..............................

10. How many acres under cultivation?..............................

11 What buildings are on the farm at present (mention value of each)?..............................

....

12. How much will it cost you to put up this building? $..............................

13. How much cash have you? $..............................

14. How much can you pay us each month? $.................

15. Give name of bank you do business with..............................

Your Name..............................

Postoffice.................................State..............................

R. F. D. No..........Box No.........Street and No..............................

PLEASE ANSWER ALL QUESTIONS.

SEARS, ROEBUCK AND CO. - - CHICAGO

Important Notice!

THE barns we offer in this catalog are designed to meet almost any demand for interior arrangement. They have been built many times over, and actual experience proves that they are correct.

For this reason you will have no difficulty in selecting a barn that meets your requirements, the interior of which can be arranged to supply your needs. Use the sketch sheet enclosed with this book to indicate what kind of floor plan you prefer if you cannot use any of the plans shown in this book.

Our Service Department is ALWAYS ready to assist you. If, during the construction of your barn, you should fail to understand some detail, which is unlikely with our proved "Already Cut" system, write us, stating just what the difficulty is. We will promptly set you right. If necessary, we will send one of our specialists out on the job to give you his personal advice. This service is FREE of charge.

The specialists of our Modern Farm Building Department are men who devote their lives exclusively to designing and constructing farm buildings. The first service they render you is to make barn building simple, and in connection with our low prices, attractive.

The much discussed "difficult task" of barn building has been reduced by us to a pleasing undertaking. This is probably the greatest service we render. All of this is covered by our liberal guarantee of "satisfaction or money back." All service is tendered for one purpose only—to help you build SUCCESSFULLY, without trouble or extra expense. It is to our personal interest to give you our BEST advice, for our reputation, our most valuable asset, is at stake in doing this.

QUALITY TALKS. The quality of our service and of our material, recommended and indorsed by you, are our only salesmen. We extend to you a cordial welcome to use the complete facilities of our Service Department.

Sears, Roebuck and Co., Chicago

COMPLETE LINE OF
HONOR-BILT
MODERN FARM
BUILDINGS
Sears, Roebuck and Co, Chicago

About the Editors

Rebecca Hunter and Dale Wolicki take a break between expeditions to search for mail order buildings. (Photo by Marilyn Raschka)

Architectural historian and researcher Rebecca Hunter became fascinated by the concept of mail order homes when she discovered the book *Houses by Mail* on a library shelf. In the hopes of getting some exercise on a snowy day in February 1997, she walked around her neighborhood looking for a home that matched one of the pictures in the book. Seventy houses later, she approached the Elgin Illinois Heritage Commission, inaugurating her first effort to locate and document these homes, which constitute a unique part of America's architectural history. Her book, *Elgin Illinois Sears House Research Project,* grew out of this intensive study.

Two years later, having discovered that "the Sears house my uncle built in Peoria" was actually a house manufactured by the Gordon-Van Tine Company in Davenport, Iowa and marketed by Montgomery Wards, she expanded her study to include homes by Gordon-Van Tine, Montgomery Ward, Harris Brothers, Aladdin, Lewis, Sterling, Pacific Homes, and Bennett companies, all of which marketed pre-cut homes on a nationwide basis during the first half of the twentieth century.

Because most mail order companies no longer have their sales records, Hunter travels to do architectural surveys to determine which homes are of possible mail order origin so that they can be investigated and authenticated. She is also available to give lectures and workshops on mail order homes and how to authenticate them.

Dale Wolicki, a native of Bay City Michigan, where the first precut homes were designed and manufactured, took an early interest in mail order homes since his father and grandfather worked in this industry. He has worked with the Bay County Michigan Historical Society to identify and locate mail order homes from Aladdin, Lewis, and Sterling companies. In addition to documenting the history of Bay City Michigan architecture, he has written a history of the Gordon-Van Tine Company of Davenport Iowa.

Hunter and Wolicki often team up and combine their expertise to search the nation's highways and byways for mail order buildings.